micro

Vieth
Membrane Systems

Wolf R. Vieth

Membrane Systems: Analysis and Design

Applications in Biotechnology, Biomedicine and Polymer Science

with 146 Figures and 37 Tables

Hanser Publishers, Munich Vienna New York

Distributed in the United States of America by
Oxford University Press, New York
and in Canada by
Oxford University Press, Canada

CHEMISTRY

Author:
Prof. Wolf R. Vieth
College of Engineering, Rutgers University, Piscataway NJ 08854

Distributed in USA by
Oxford University Press
200 Madison Avenue, New York, N. Y. 10016

Distributed in Canada by
Oxford University Press, Canada
70 Wynford Drive, Don Mills, Ontario, M3C IJ9

Distributed in all other countries by
Carl Hanser Verlag
Kolbergerstraße 22
D-8000 München 80

The use of general descriptive names, trademarks, etc., in this publication, even if the former are not especially identified, is not to be taken as a sign that such names, as understood by the Trade Marks and Merchandise Marks Act, may accordingly be used freely by anyone.

While the advice and information in this book are believed to be true and accurate at the date of going to press, neither the authors nor the editors nor the publisher can accept any legal responsibility for any errors or omissions that may be made. The publisher makes no warranty, express or implied, with respect to the material contained herein.

CIP-Titelaufnahme der Deutschen Bibliothek

Vieth, Wolf R.:
Membrane systems: analysis and design : applications in
biotechnolgy, biomedicine, and polymer science / Wolf R.
Vieth. – Munich ; Vienna ; New York : Hanser; New York :
Oxford Univ. Press, 1988
  ISBN 3-446-15206-7 (Hanser) Pp.
  ISBN 0-19-520767-X (Oxford Univ. Pr.) Pp.

ISBN 3-446-15206-7 Carl Hanser Verlag Munich Vienna New York
ISBN 0-19-520767-X Oxford University Press

DEDICATION

This book is dedicated to my beloved wife, Peggy, and to my family, the well-springs of my inspiration and encouragement. Along with her love and support, this project could not have succeeded without Peggy's excellent word-processing, editing and sketching efforts.

PREFACE

The science and engineering of membrane systems are rapidly becoming of major interest as applications arise in such diverse fields as Biotechnology, Biomedicine and Polymer Science. There is a definite need to pull together the seemingly disparate threads of knowledge which are appearing in the several disciplines.

The writing of this book arose out of these convictions, as well as from the perspective of more than twenty five years of research and teaching in these areas. Quite naturally, in attempting such a synthesis of ideas, I rely rather strongly on work carried out in our own laboratory in pursuit of the goal of unification. The purpose of this book, then, is to unify the principles of diffusion and reaction which are applying at the molecular level in synthetic and biological membrane systems. This is an ambitious task and the writer freely acknowledges that this work is just a beginning; it makes no pretense of setting out to be all-encompassing. But it is a necessary and worthwhile beginning, I believe.

While it is true that the subjects of molecular kinetics, diffusion and thermodynamics are in relatively good shape (Sherwood et al., 1975), their interrelation with membrane (polymeric) microstructure adds a new and exciting dimension. By the same token, chemical and physical phenomena are displayed in a particularly simple, often unidimensional domain, where their analysis is tractable and revealing.

It is especially rewarding to explore the interactive control systems encountered in biology. The inherited wisdom of evolution is on display whereby efficient use of relatively few elements; i.e., simple chemical messengers interacting with complex biopolymers in membrane structures is revealed. And yet, the function of these highly sophisticated systems is reflected in the behavior of simpler synthetic membrane systems, as well as in the design of artificial ones; for example, asymmetric enzyme membranes, as biosensors. The common theme is the *response of a membrane system to penetrant-induced conformational changes* ; frequently, this response is multimodal, as we shall see.

At this point, it is of interest to define more precisely what is meant by some of the terms I am employing. By system, I mean "a regularly interacting or interdependent group of items forming a unified whole." The lac operon is a good example of a system, albeit as part of a still larger one (the synthetic machinery for ß- galactosidase). I will employ a process approach; that is, "a series of actions or operations conducing to an end," literally taking a system apart (analysis) to see "what makes it tick," prior to recomposing it (synthesis or design).

The approach just outlined is readily identified as one which originated in Cartesian thinking. To Dean Elmer Easton I owe another debt, for his definition of Engineering which is, "The humane art and science of employing knowledge of the materials and energy of nature for the creation of utility and beauty." In this spirit, I liken the approach I share with kindred research colleagues to a kind of mental sculpture; that is, the shaping of ideas into the images of phenomena and processes. Thanks to many such efforts, some of these images have been polished to a certain smoothness and lustre; for instance, dual sorption theory. Others might still have a few rough spots; for example, transport regulation of anaerobic processes. Lastly, the reader will recognize that all the foregoing is rooted in the continuing intellectual traditions which began in the Renaissance, a modern chapter of which I believe is being written even now in our profession.

ACKNOWLEDGEMENTS

A man's character is shaped through his experiences; two periods which had considerable influence for me were those spent as Director of the MIT School of Chemical Engineering Practice in the 1960s and as a Department Chairman during the formative years for Rutgers Biochemical Engineering in the 1970s. Prior to those days, I was fortunate to acquire a firm foundation in Chemical Engineering and Applied Chemistry under the tutelage of such Professors as C.E. Dryden, T.K. Sherwood and A.S. Michaels. To them and to my close colleagues, Drs. K. Venkat, G.K. Chotani, S. Hirose, S. Gondo, T. Matsuura and K. Sladek, I owe a special debt of thanks.

I am grateful to the administrative officers at Rutgers, especially Drs. E.J. Bloustein, T.A. Pond and E.H. Dill, for their support and for approving my application for Sabbatical leave to write this book. Likewise, I appreciate the contributions and/or constructive criticism of my faculty colleagues at Rutgers: Profs. Alkis Constantinides, Henrik Pedersen, John Sauer, Darrell Morrow, Seymour Gilbert, Kan-Ichi Hayakawa, Shaw Wang and Burton Davidson. I have learned much, too, from my professional colleagues: Drs. Shuichi Suzuki, Isao Karube, Walter Marconi, Tarun Ghose, T.M.S. Chang, Csaba Horvath and Don Paul, Harry Gregor, Vivian Stannett, Harold Hopfenberg, Norman Li, Alex Stern, Jim Barrie, Enrico Drioli, Bill Koros and Harry Frisch, John Petropoulos, Joop Roels, Jay Bailey, Mike Shuler, Bob Tanner, Bill Weigand, Don Kirwan, Bob Coughlin, Henry Lester and many others. All of these individuals helped me to gradually acquire sharper tools.

It is a distinct pleasure to acknowledge the work of my students and postdoctoral fellows; their contributions, marked by reference citations, illuminate the pages of this book. I wish to thank especially Kwang Nho who prepared the graphs which appear in each chapter and Nick Bosko, a true friend in need in the laboratory and outside. I am grateful to Dr. E. Immergut and his colleagues at Hanser Publishers for their assistance and unfailing courtesy.

All in all, the experiences leading up to this book as well as its actual preparation were for me a voyage of discovery, full of wonders and marvels; I would not have traded it for anything. Perhaps I can express this feeling more eloquently through the idiom of folk music, another passion of mine, by introducing each main chapter with a few phrases from pieces which have special meaning for me.

I wish to acknowledge the permission of the authors or publishers of the following journals or books to reproduce the figures and portions of text specified: Figs. 2.4, 2.11, 3.3, Koros, W.J., *J. Polym. Sci.*, J. Wiley and Sons Inc.;   Fig. 2.5, Koros, W.J., *J. Mem. Sci.*, Elsevier/North Holland Inc.;   Figs. 3.5, 3.6, Paul, D.R., *J. Appl. Polym. Sci.*, John Wiley and Sons;   Fig. 3.7, Toi, K., *J. Polym. Sci.*, John Wiley and Sons;   Fig. 3.12, Stern, S.A., *J. Mem. Sci.*, Elsevier/North Holland;   Fig. 3.13, Berens, A.R., *J. Mem. Sci.*, Elsevier/North Holland;   Fig. 3.16, Paul, D.R., *J. Mem. Sci.*, Elsevier/ North Holland;   Portions of text (including tables), Ch.5, pp. 132 through 145, *Appl. Biochemistry and Bioeng.*, Academic Press Inc.;   Ch.2, p. 19; Ch.3, pp. 79-83, J. *Mem. Sci.*, Elsevier/North Holland;   Ch.7, pp. 212-224, 227-231 and Ch.9, pp. 299-310, *J. Mol. Cat.*, Elsevier Sequoia S.A.

Wolf Vieth

Cape May Point, New Jersey
August, 1987

CONTENTS

Gaudeamus Igitur,
Juvenesdum Suumus!

Trad. Academic Song

1

INTRODUCTION

1.0 GENERAL

The diffusive transport of matter across a membrane in response to an activity gradient is a unifying thread in the fabric of the chapters which are to follow, so a brief "primer," so to speak, is in order. In the simplest case, molecular transport occurs via a random walk mechanism which concludes with desorption of the penetrant from the surface at the lower concentration. The total permeation process consists then of sorption, diffusion and desorption.

With a time-invariant concentration difference across the membrane (see Figure 1.1) the steady state, unidirectional flux of gas can be described by Fick's first law of diffusion:

$$J = -D \frac{\partial c}{\partial x} \qquad\qquad [1.1]$$

where J is the flux and $\partial c/\partial x$ the concentration gradient. (When a counterposing gradient involving another chemical component exists, one is dealing with the process of *counterdiffusion*.)

To consider a simple standard case, the sorption of gas in rubbery polymeric membranes is so low that gas-gas interactions are negligible and D is independent of concentration.

Thus with $D \neq f(c)$ $\qquad\qquad\qquad\qquad$ [1.2]

$$J = D [c_1 - c_2] / \ell \qquad\qquad [1.3]$$

where c_1 and c_2 are the concentrations at the upstream and downstream surfaces of the membrane, respectively, and ℓ is the thickness of the membrane.

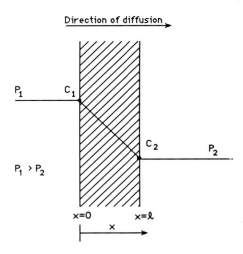

Figure 1.1   Molecular diffusion across a polymer membrane.

Accumulated evidence has demonstrated that sorption of gases in rubbery polymers is often well-described by Henry's Law, so that the concentration of the gas, c, at either surface of the membrane can be related to its partial pressure there:

$$c = kp \qquad\qquad [1.4]$$

Thus,

$$J = Dk \frac{[p_1 - p_2]}{l} \qquad\qquad [1.5]$$

where k is the Henry's Law solubility constant, and p_1 and p_2 refer to the pressures at the film surfaces.

By definition, the product of D and k is $\overline{P}$, the permeability constant,

$$\overline{P} \equiv kD \qquad\qquad [1.6]$$

and,

$$\overline{P} = \frac{J \, l}{[p_1 - p_2]} = \frac{\Delta Q}{\Delta t} \frac{l}{A \, [p_1 - p_2]} \qquad [1.7]$$

where ΔQ is the amount of gas transmitted in the given interval of time, Δt, and A is the area of the membrane exposed to the diffusing gas.

1.1 TRANSIENT STUDIES

Much progress has been made by focusing attention on separate measurement of the diffusion and solubility constants and correlating the data obtained according to molecular models of the permeation process. An important step was taken by Barrer (1939) who adapted the dynamic method of Daynes (1920), called the time lag method, to the measurement of diffusion constants. The mathematical basis for this procedure is an integration of Fick's second law,

$$D \frac{\partial^2 c}{\partial x^2} = \frac{\partial c}{\partial t} \qquad [1.8]$$

applying to unsteady state, unidirectional diffusion. The necessary boundary conditions are an initially gas free film, attainment of equilibrium at the inlet gas-polymer interface, and zero concentration of gas held at the polymer outflow face. The first condition is assured by evacuation of the film prior to introduction of the permeant, the second has been experimentally found to apply for many polymer-gas systems, through sorption isotherms, and the third condition is automatically satisfied because of the extremely slow rates of permeation encountered experimentally. These conditions being satisfied, extrapolation to zero pressure of the steady state portion of a plot of pressure downstream of the polymer membrane versus time yields a value θ, called the time lag. This extrapolation may be accurately accomplished because the pressure-time plot is linear after steady state has been attained. The time lag is simply related to the diffusion constant.

$$D = \frac{l^2}{6\theta} \qquad [1.9]$$

Permeabilities can be calculated from the amount of gas which has passed through the polymer in the course of an experiment. The gas is collected and application of the perfect gas law allows a calculation of $\overline{P}$. From the ratio of $\overline{P}/D$, the solubility constant k is determined.

1.2 MICROHETEROGENEITY IN MEMBRANES

In contrast to the "ideal" case of the rubbery polymer membrane, synthetic polymer membranes and biological membranes frequently display various kinds and degrees of microheterogeneity. In the case of synthetic glassy polymers, this effect arises out of the "freezing-in" of unrelaxed excess free volume as the polymer passes through the glass transition temperature, below which segmental translations and rotations do not readily occur. The resulting structure contains penetrant-entrapping microvoids, as well as regions which are rich in amorphous chain segments of more normal density. For biological membranes, the coexistence of lipids and proteins in the structure produces the effect. In both cases, the presence of microheterogeneity confers a multi-fold transport character on the structure. For penetrants in glassy polymers, duality of transport modes often suffices to explain the results. For biological membrane systems, various hierarchies are possible, including the duality of passive and active transport.

The interactions of the penetrant with the polymer microstructure are conveniently described with equilibrium thermodynamic models; Langmuir and Henry's law relations for synthetic systems and ligand binding relations such as the Monod allosteric model for biological membrane systems. In the latter case, the model can be described as a Langmuir relation with a variable site saturation coefficient.

In all these instances, the influences of the penetrant binding reactions on the diffusional process can be profound, particularly in the transient regime. Reversible site-binding reactions almost invariably prolong the time lag which is a measure of the time required to reach the stationary state for a membrane system. These delays are offset in biological systems through the agency of essentially irreversible forward reaction processes which produce sharper gradients.

In order to comprehend these diverse phenomena it is perhaps best to cut one's teeth (as did the author) on the synthetic membrane system where the structure may be invariant   under challenge by a penetrant.   Next, one can move on to a structure which is a hybrid synthetic polymer-biopolymer membrane system; i.e., an enzyme membrane or biosensor.   Then, membrane enclosure of active biological entities such as enzymes or whole cells leads logically to a consideration of biocatalysts and bioreactors.   These chapters provide a bridge between the purely synthetic membrane systems treated near the beginning of this book and the purely biological membrane systems treated near its conclusion.

1.3 BIOSENSORS AND ENZYME CATALYSIS

It is helpful to begin a discussion of biological membrane systems with the consideration of enzyme biosensors.   Now, to the elementary transport steps of sorption and diffusion for a penetrant, we must add the elementary reaction steps which consume the penetrant or produce it from another species, the "substrate."

KINETIC BEHAVIOR OF IMMOBILIZED ENZYMES

The simplest model of conversion of substrate (S) to product (P) catalyzed by unsupported enzyme (E) is,

$$E + S \underset{k_{-1}}{\overset{k_1}{\Leftrightarrow}} ES \overset{k_2}{\Rightarrow} P + E$$

where k_1, k_{-1} and k_2 are kinetic constants.   If one assumes that a steady state exists in which the concentration of intermediate (ES) does not vary with time, the Michaelis-Menten relationship can be developed:

$$V = \frac{k_2 E_O S}{K_m + S} \tag{1.10}$$

where V = velocity of the enzyme reaction; K_m = Michaelis-Menten constant; E_O = total enzyme concentration; and S = substrate concentration.

If the concentration of substrate is large relative to K_m, then $V = k_2 E_O = V_m$ and the rate of reaction is at its maximum.

When an enzyme is attached to a solid support, the kinetic pattern of reaction changes considerably, leading to changes in the values of the kinetic parameters K_m and V_m.   The kinetics of such reactions are obscured by several factors and the observed kinetic parameters are only "apparent." These factors are: (a) change in enzyme conformation, (b) steric effects, (c) microenvironmental effects, and (d) bulk and internal diffusional effects.

DIFFUSIONAL LIMITATIONS

Unlike soluble enzyme, matrix-supported enzyme has to exercise its catalytic action in a heterogeneous environment. As in conventional heterogeneous catalysis, at least five distinct steps in the overall enzymatic process can be identified: (1) diffusion of the substrate from the bulk phase to the carrier surface, (2) transport of the substrate from the carrier surface to the domain of the enzyme, (3) enzyme-catalyzed conversion of substrate, and (4) reversal of steps (2) and (1) for product. Steps (l) and (5) are external or bulk diffusional effects, while steps (2) and (4) are internal or pore diffusional effects.

Detailed models taking into account simultaneous mass transfer and biochemical reaction are then developed. In addition, electrostatic effects must be incorporated quantitatively wherever necessary.

1.4 BIOCATALYSTS AND BIOREACTORS

As a prelude, enzyme membranes and microbial membranes employed in biosensors are discussed. The interactions of membrane diffusion and reaction and their effects on sensor time lags are analyzed; improved designs such as anisotropic enzyme membranes are presented.

Next, we go on to examine what occurs when biocatalytic structures of the type considered above are placed into efficient bioreactor-separator configurations. The reactor balance equations make their appearances naturally, and the previously derived diffusion-kinetic models are incorporated. The contributions of our laboratory to the establishment of a biocatalyst-bioreactor technology based on the biopolymer, *collagen* are likewise detailed.

1.5 CHEMICAL MESSENGERS

The penetrant-microstructure interactions described above (e.g. biosensors), amount to a simple chemical message transport system in a control theory sense. Biological scientists have long been fascinated with such control and/or sensing systems which actually occur in nature, both inter- and intra-cellularly. The lac operon is an excellent example of the former case and the role of neurotransmitters at the neuromuscular junction, the latter. The lac operon is under genetic control and the description of immobilized recombinant (lac) cell reactors is a natural extension.

The post-synapse comprises a collagen-immobilized enzyme system as a support structure for the post-synaptic membrane. Recognition of this leads to the formulation of a diffusion-kinetic model for events at the neuromuscular junction where penetrant duality reappears. The book then concludes with a description of mammalian excitable membranes, those of the myocardium in particular.

Our objectives in the ongoing research leading up to this book are:

i. To construct a comprehensive set of principles of molecular diffusion and chemical reaction to describe the penetrant-immobilizing, glassy-state behavior of synthetic polymers;

ii. Using these principles as a guide, to generalize and extend the resulting diffusion-kinetic models for microheterogeneous polymeric systems to the realms of biology and, of course, biopolymers, to enable analysis and design of biosensors, biocatalysts and bioreactors;

iii. To proceed to a consideration of living systems, per se, emphasizing the general importance in biotechnology of cell membrane-permeating chemical messengers under genetic control;   and, in a parallel vein,

iv. To generalize biosensor theory to encompass analysis of messenger-induced transitions of biopolymeric structures involved in neuromuscular responses and to enable design of advanced membrane biosensor systems.

We hope that a journey through this book will be a rewarding one for the reader and that it furnishes at least some measure of the excitement we have experienced along the way!

1.6 REFERENCES

Barrer, R.M., *Trans Farad. Soc.*, **35**, 628 (1939).
Daynes, H.A., *Proc. Roy. Soc. Lond.*, Ser A **97**, 286 (1920).

The fox went out on a chilly night,
Prayed to the moon for to give him
     light,
For he'd many a mile to go that
     night,
Afore he reached the town-oh.

Folk Ballad, "The Fox"

DIFFUSION AND REACTION
IN MICROHETEROGENEOUS SYNTHETIC MEMBRANES

2.0 RUBBERY POLYMER MEMBRANES

We begin, then, with an in-depth examination of small molecule transport in polymer membranes which will occupy us for the next two chapters of this book. For the rather straightforward case of rubbery polymer membranes, as described already in Chapter 1:

$$J = -Dk \frac{dp}{dx} \qquad [2.1]$$

By substitution,

$$J = -\overline{P} \frac{dp}{dx} \qquad [2.2]$$

The temperature dependencies of the three parameters are then readily expressed, as:

$$D = D_0 \exp[-E_D/RT] \qquad [2.3]$$

$$k = k_0 \exp[-\Delta H_s/RT] \qquad [2.4]$$

$$\overline{P} = \overline{P}_0 \exp[-E_{\overline{P}}/RT] \qquad [2.5]$$

2.1 PENETRANT LOCALIZATION

The concept that a second mechanism of sorption may be implicated in the solution and diffusion behavior of small molecules in amorphous polymers originated with Meares (1954).   His investigations indicated that the glassy state contains a distribution of microvoids frozen into the structure as the polymer is cooled through its glass transition temperature.   Free segmental rotations of the polymer chains are restricted in the glassy state, resulting in fixed microvoids or "holes" throughout the polymer.   These microvoids in the glassy polymer network act to immobilize a portion of the penetrant molecules by entrapment or by binding at high energy sites at the molecular periphery of the microvoids (similar to adsorption).

Based on Meares' concept of microvoids in the glassy state and the observation that sorption isotherms of organic vapors in ethyl cellulose showed characteristic curvature concave to the pressure axis, Barrer et al. (1958) suggested two concurrent mechanisms of sorption: ordinary dissolution and "hole-filling."

Subsequent research (Assink, 1975; Koros, 1980; Koros et al., 1981; Michaels et al., 1963a, b; Vieth, 1961; Eilenberg and Vieth, 1972; Vieth and Sladek, 1965) has demonstrated that the sorption isotherms of small molecule penetrant gases such as carbon dioxide, methane, argon and nitrogen in glassy polymers are generally concave to the pressure axis.   The dual mode sorption model is now widely used to describe such behavior.   This model describes the sorption mechanism in terms of one population of ordinarily dissolved sorbate which resides within the polymer matrix and is described by Henry's law, while the second population of sorbate is considered to occupy unrelaxed free volume within the polymer and is described by a Langmuir isotherm.

The equilibrium part of the model is simply expressed by the following equation for the isotherm:

$$C = C_D + C_H = k_D p + \frac{C_H' bp}{1 + bp} \qquad [2.6]$$

where C = solubility, cc (STP)/cc polymer; k_D = Henry's law dissolution constant, cc(STP)/cc polymer atm; b = hole affinity constant, atm^{-1}; C_H' = hole saturation constant, cc(STP)/cc polymer; and p = pressure, atm. The first term, C_D, represents sorption of normally diffusible species while the second term, C_H, represents sorption in microvoids or "holes." The hole affinity constant, b, is related to the ratio of rate constants of sorption and desorption of penetrant in the holes.   Where bp $\ll$ 1, the sorption isotherm, eqn. [2.6], reduces to a linear form:

$$C = [k_D + C_H' b] p \qquad [2.7]$$

At sufficiently high pressures, the microvoids become saturated and will no longer sorb additional penetrant. When $bp \gg 1$, sorption in the microvoids reaches the saturation limit, C_H' , and eqn. [2.6] again reduces to a linear form:

$$C = k_D p + C_H' \qquad\qquad [2.8]$$

Thus, the dual sorption model predicts that an isothermal plot of C vs. p will consist of a low-pressure linear region and a high-pressure linear region connected by a nonlinear region. The sorption isotherms for methane in polystyrene shown in Fig. 2.1 indicate that this type of sorption behavior does indeed occur.

It is possible to quantitatively separate the two contributions to sorption. As shown by eqn. [2.8], k_D can be obtained from the slope of the sorption isotherm at high pressures. By subtracting the solubility of the normally dissolved gas, C_D, from the total solubility C, the gas solubility in the microvoids, C_H, can be calculated at each pressure:

$$C_H = C - C_D = C - k_D p \qquad\qquad [2.9]$$

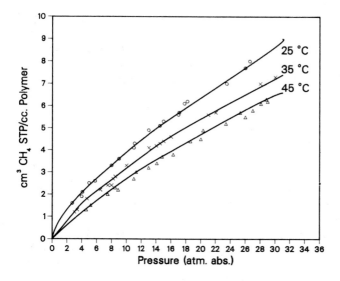

Figure 2.1   Solubility of methane in oriented polystyrene.

Once C_H as a function of p has been determined, it can be tested in terms of the Langmuir model:

$$C_H = \frac{C_H'bp}{1 + bp} \qquad [2.10]$$

Equation [2.10] can be rearranged to:

$$\frac{p}{C_H} = \frac{1}{C_H'b} + \frac{p}{C_H'} \qquad [2.11]$$

Therefore a test of whether a single Langmuir expression fits the data for the hole-filling process is to make a least-mean-squares plot of p/C_H vs. p and to check for linearity. The constants C_H' and b can be determined from the slope and y-intercept of such a Langmuir plot.

Analysis of sorption data in terms of the dual sorption model permits a delineation of the temperature dependencies of the individual, uncoupled sorption parameters. Using the van't Hoff expression:

$$k_D = k_{D_0} e^{-\Delta H_D/RT} \qquad [2.12]$$

the enthalpy of dissolution (ΔH_D) can be determined from the slope of a plot of $\ln k_D$ vs. $1/T$. Similarly, from the temperature dependency of the hole-affinity constant, b, the apparent enthalpy of hole-filling can be estimated. In this manner, anomalously high negative enthalpies of sorption in glassy polymers can be rationally explained by accounting for a second mode of sorption.

For example, as shown in Fig. 2.2, the overall sorption enthalpy for methane in glassy polystyrene has been found to be -6.2 kcal/g mole (Vieth et al., 1966), much higher than the value characteristic of normal dissolution in rubbery polymers. Analysis of the sorption data in terms of the dual sorption theory, however, yields a normal enthalpy of dissolution of -1.5 kcal/g mole, which is more consistent with values observed in rubbery polymers. The enthalpy of hole-filling is -3.6 kcal/g mole, indicating the proportionately large contribution of this second mode of sorption to the overall apparent enthalpy of sorption. The hole saturation constant, C_H', varies from 3.25 cc(STP)/cc polymer

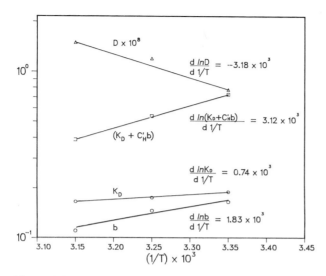

Figure 2.2   Sorption parameters and diffusion coefficients, vs. 1/T for methane in oriented polystyrene.

to 1.98 cc(STP)/cc as the temperature is raised from 25°C to 45°C. The activation energy for the mobile species (6.4 kcal g mole^{-1}) is correspondingly lower than the apparent one (11.1 kcal g mole^{-1}), based on total concentration (Barrie et al., 1975).   Thus, when the sorption parameters are uncoupled and the enthalpy contributions of each mode separately determined, the data seem to make much more sense and are in closer agreement with what is known about the transport process in rubbery polymers.   Additionally, the temperature dependency of $C_H{}'$, first noted in this research, suggests the possibility of tracking excess volume changes through application of sorption technique and analysis.

Based on the simple dual-mode treatment, Koros extended the theory to sorption of gas mixtures in glassy polymers (1980).   He first assumed that the primary effect for a mixture is competition by the various penetrants for the fixed unrelaxed free volume in the polymer. His second assumption is that, in the Henry's law region, solubility of a given penetrant is essentially independent of the other components present.   The sorption equilibrium isotherms representing the binary mixture of gases A and B are:

For Gas A:

$$C_A = C_{DA} + C_{HA} \qquad\qquad [2.13]$$

$$C_A = k_{DA}P_A + \frac{C'_{HA}b_AP_A}{1 + b_AP_A + b_BP_B} \qquad [2.14]$$

For Gas B:
$$C_B = C_{DB} + C_{HB} \qquad [2.15]$$

$$C_B = k_{DB}P_B + \frac{C'_{HB}b_BP_B}{1 + b_AP_A + b_BP_B} \qquad [2.16]$$

Here, C_i is the total concentration of component i, C_{Di} is the concentration in the Henry's law region and C_{Hi} is the concentration in the Langmuirian region. For a binary mixture, A and B refer to the components, k_{DA} is the Henry's law parameter for component A, b_A is the Langmuir affinity of component A, C'_{HA} is the Langmuir saturation parameter, and P_A is the partial pressure of component A.

All the parameters in the above equations can be obtained either by combinations of single gas sorption measurements or by mixed gas sorption measurements. Either way the values are found to be the same (Sanders, Koros, Hopfenberg and Stannett, 1984). This is a powerful advantage of the dual mode sorption model: once the parameters of single gases are known, equations [2.14] and [2.16] can predict the sorption equilibrium behavior of mixed gases. Experimental confirmation of this model has been carried out by Sanders et al. (1983), Chern et al. (1983), Sanders et al. (1984) and Sanders and Koros (1986). They measured the sorption of CO_2 and C_2H_4 and mixtures of these gases in poly(methyl-methacrylate) at 35°C. The total solubility of each component in the mixture is reduced when compared to that for the corresponding single component at the same partial pressure. The reduction results from the competition for Langmuirian sites and is accurately described by the dual mode sorption model. The authors framed the issue that permeability reduction for mixed gases which was reported in other work (Pye et al., 1976a, b; McCandless, 1972; Chern et al., 1982) may result from either the solubility effect or the dif-fusivity effect. Based on their sorption work, the authors concluded that permeability reduction of each component in mixed-gas sorption results from the solubility effect. This implies that the intrinsic diffusivity of a gas in the polymer is not changed by introducing a second gas.

2.2 GLASSY POLYMERS AND DUAL MODE TRANSPORT

The formulation of the partial immobilization model and the mixed gas isotherms leads to transport equations which will be presented later.   Analysis begins here by considering the early development known as the total immobilization model.

The author and Karl Sladek (1965) developed a model which describes the transient sorption of penetrant in a polymer when dual modes of sorption occur - the kinetic part of the original dual sorption theory.   The basic assumptions applied to the transport model are:

1. Two modes of sorption, Henry's law sorption and Langmuir sorption, occur simultaneously.
2. Local equilibrium between the two modes is maintained throughout the membrane.
3. The gas sorbed in the Langmuir mode is completely immobilized.
4. Diffusion occurs only in the Henry's mode.
5. The true diffusion coefficient is a constant, independent of concentration or position in the membrane.

With the above postulations, the unsteady state transport model was derived for single gas diffusion.   According to Fick's first law of diffusion:

$$J = - D_D \frac{\partial C_D}{\partial x} \qquad [2.17]$$

where D_D is the diffusion coefficient of the Henrian species.   For unsteady state, unidirectional transport of gas:

$$\frac{\partial}{\partial t} [C_D + C_H] = - \frac{\partial J}{\partial x} \qquad [2.18]$$

One substitutes eqn. [2.17] into eqn. [2.18] to obtain:

$$\frac{\partial}{\partial t} [C_D + C_H] = D_D \frac{\partial^2 C_D}{\partial x^2} \qquad [2.19]$$

Since local equilibrium exists throughout the membrane, C_H can be eliminated by equating the partial pressures P in eqn. [1.12] to yield:

$$D_D \frac{\partial^2 C_D}{\partial x^2} = \frac{\partial C_D}{\partial t} [1 + (C'_H b k_D^{-1} / [1 + [b/k_D] C_D]^2)] \qquad [2.20]$$

By a numerical method a general solution was obtained for this model for any set of parameters C'_H, b and k_D. Curve matching of sorption-kinetic data permitted specifying the diffusion coefficient (Fig. 2.3). Linearized versions of eqn. [2.20] had been presented earlier (Crank, 1956; Michaels et al., 1963). Eilenberg and Vieth (1972) later researched sorption kinetics and equilibria for CO_2 in polycarbonate. Fenelon used this method to predict the barrier properties of Mylar® polyester film (1973).

By using the method of Frisch (1957), Paul developed the time lag function for the total immobilization model (1969). He found that the penetrant time lag depends on the boundary concentration used in permeation. When gas molecules diffuse through a membrane, part of the population is diverted to fill unrelaxed free volume, appearing as microvoids, thus delaying the diffusion process and increasing the time lag. As the boundary concentration increases, the driving force is increased and the diffusing gas is better able to fill these holes, which results in a decrease of the time lag. Thus, time lag decreases as boundary pressure increases.

Early transport research concentrated on the kinetics of high pressure sorption. To determine the transport parameters from transient permeation remained a further goal. Measurement of time lags would provide a rigorous test of dual sorption theory, it was felt, and this belief turned out to be accurate.

The possibility that gas molecules sorbed in the Langmuirian mode may not be completely immobilized was noted by Petropoulous (1970) and Paul (1969). Petropoulous eliminated the complete immobilization assumption and suggested a general transport process achieved by separation of the effects for each of the two modes. Assuming that the chemical potential gradient is the driving force for diffusion, the unsteady state equation can be written as:

$$\frac{\partial C}{\partial t} = \frac{\partial}{\partial x} [(R_g T)^{-1} (D_{T1} C_D + D_{T2} C_H) \partial \mu / \partial x] \qquad [2.21]$$

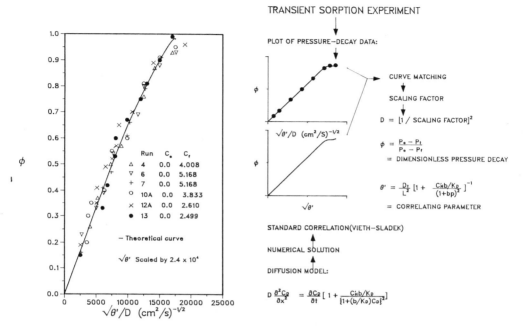

Figure 2.3 Estimation of diffusivities by the method of Vieth and Sladek (1965).

where μ represents the chemical potential, R_g is the gas constant and T is the absolute temperature. D_{T1} and D_{T2} are the thermodynamic diffusion coefficients of dissolved and adsorbed penetrant, respectively. This approach has subsequently been applied successfully to dye diffusion in polymer matrices (Komiyama and Iijima, 1974).

Paul and Koros (1976) extended Petropoulous' work by using concentration as the driving force for diffusion. The resulting model is known as the dual mobility or partial immobilization model. They suggested that the Langmuir population is not completely immobilized but has a partial mobility. In their model, the dual mobility transport of a gas in a glassy polymer is described by Fick's law:

$$J = J_D + J_H$$

$$= -D_D \frac{\partial C_D}{\partial x} - D_H \frac{\partial C_H}{\partial x} \qquad [2.22]$$

where J is the total diffusional flux, J_D and J_H are the respective fluxes of the two populations, C_D and C_H, described in eqn. [2.12], and D_D and D_H are the respective diffusion coefficients of the two sorbed populations. Another approach for penetrant transport is to define a phenomenological diffusion coefficient as the "effective diffusivity," which expresses the flux in terms of the total concentration, C (Koros et al., 1976).

$$J = - D_{eff} \frac{\partial C}{\partial x} \qquad [2.23]$$

The difference between the two approaches is that eqn. [2.22] takes account of the unique character of gas sorption in glassy polymers and yields parameters with the potential for physical interpretation, while eqn. [2.23] does not readily permit such an approach. The permeability of pure gases (Paul and Koros, 1976) predicted from the partial immobilization model measured with an upstream pressure, p, and zero downstream concentration is given by:

$$\overline{P} = k_D \, D_D \, [\, 1 + (FK \, / \, 1 + bp) \,] \qquad [2.24]$$

where $F = D_H/D_D$ and $K = C'_H b/k_D$. The time lag (Paul and Koros, 1976) predicted from the partial immobilization model yields a function depending on upstream pressure and sorption constants. Typical results are shown in Fig. 2.4. Alternative models have recently appeared (Raucher and Sefcik, 1983; Chern et al., 1983). Interpretation of some of the parameters has been modified, in the work of Frederickson and Helfand (1985).

Koros, Chan and Paul have demonstrated the applicability of a dual sorption/dual mobility model in the transport of gases in polycarbonate (1977) and Koros and Paul, carbon dioxide in poly(ethylene terephthalate) (1978a,b). They found that permeability and time lag are predictable functions of boundary pressure. In their work, they combined steady state permeation and sorption data to specify all five parameters. The parameters k_D, b and C'_H can be determined by fitting an isotherm equation to sorption data. A least mean squares fit of the steady state permeability data versus the dimensionless variable $1/(1+bP)$ yields D_D and F from the intercept and slope, respectively. The adequacy of the partial immobilization model to describe the transport process was then tested by comparing the predicted time lag with the experimentally measured value. Good agreement was observed.

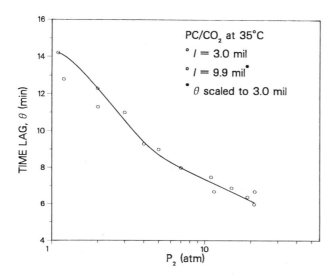

Figure 2.4 Pressure variation of time lags (Koros et al., 1976).

Tshudy and von Frankenberg (1973) considered an invariant set of N_0 immobilizing sites distributed throughout the medium, each site capable of immobilizing one penetrant molecule.

$$k_1$$
$$\rightarrow$$
Hole or Site + Penetrant Molecule $\leftarrow$ Hole - Molecule Complex
$$k_{-1}$$

The diffusion-kinetic equation for the change of penetrant concentration with time is then written as:

$$\frac{\partial C_D}{\partial t} = \frac{\partial}{\partial x} [D \cdot \partial C_D / \partial x] - k_1 \cdot n_t \cdot C_D + k_{-1} \cdot [n_0 - n_t] \qquad [2.25]$$

The microvoid saturation constant, C_H' , is analogous to the total number of microvoids in the system, N_0 , and the hole affinity constant, b, is equivalent to the ratio of $k_1 \cdot k_D / k_{-1}$. This time lag equation is equivalent to an earlier one developed by Paul (1969); θ approaches the familiar limit, $l^2 / 6D$, when no immobilization occurs.

Using the dual mode sorption isotherm for mixed gases as given by eqns. [2.14] and [2.16], Koros et al. extended the permeability model to steady state *cocurrent* permeation of a binary mixture (1980). For gases A and B, with upstream pressures P_A and P_B and zero downstream pressure, the permeability expressions for cocurrent diffusion are:

For gas A:

$$\overline{P}_A = k_{DA} D_{DA} [1 + (F_A C'_{HA} b_A P_A / 1 + b_A P_A + b_B P_B)] \quad [2.26]$$

For gas B:

$$\overline{P}_B = k_{DB} D_{DB} [1 + (F_B C'_{HB} b_B P_B / 1 + b_A P_A + b_B P_B)] \quad [2.27]$$

where $F_A = D_{HA} / D_{DA}$; $F_B = D_{HB} / D_{DB}$. Typical results are shown in Fig. 2.5.

Comparing eqn. [2.26] with eqn. [2.24], the reduction of permeability originates with the $b_B P_B$ term in the denominator which accounts for the competition for unrelaxed free volume. Performing a mixed gas cocurrent diffusion experiment provides a rigorous test of the partial mobility model. It might be argued that a counterdiffusion experiment provides an even more critical test; e.g., it is found that the transient flux may not follow the usual behavior where the flux monotonically increases from a zero level to a steady value, as with a single component. Counterdiffusion flux for the gas with the higher diffusivity can exhibit a maximum value, as will be shown.

2.3 DIFFUSION MECHANISMS

Chern et al. (1983) described four kinds of unit diffusion step for the dual mode transport of gases in a glassy polymer. These authors and Frisch and Stern (1983) all noted the possibility of factoring the diffusion coefficient into a product of a mean-square "jump" length, a jump frequency factor, and a correlation probability factor. A quantitative description of this diffusion mechanism was carried out by Barrer (1984). For the penetrants existing in the polymer, Barrer designates "D" to indicate gas molecules in the Henry's mode and "H" to indicate them in the Langmuir mode. The unit diffusion steps are then:

$$D \rightarrow D$$
$$D \rightarrow H$$
$$H \rightarrow D$$
$$H \rightarrow H$$

Each of these diffusional steps displays a characteristic diffusion coefficient: D_{DD}, D_{DH}, D_{HD} and D_{HH} respectively. Consider the flow across a unit area of any plane XY normal to the direction of flow; for each type of flow the flux is the difference of numbers of molecules

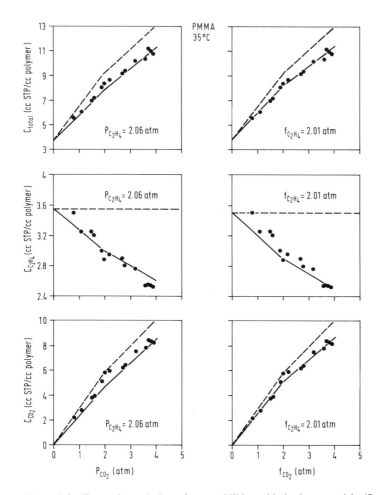

Figure 2.5   Tests of correlation of permeabilities with isotherm models (Sanders et al., 1984).

jumping left to right and right to left per unit time across unit area of the XY plane. Then the flux is the product of the molecular jump frequency and the concentration of dissolved molecules per unit area. For the type of jumps from D to D, the net flow across unit area of plane XY is:

$$J_{DD} = k_{DD} (n_{DD} - n'_{DD})$$ [2.28]

where k_{DD} is the jump frequency and the n_{DD} is the number of dissolved molecules per unit area to the left of XY. The number of molecules to the right of XY can be written as:

$$n_{DD} = n'_{DD} + \frac{dn_{DD}}{dx} \, \mathsf{L}_{DD} \qquad\qquad [2.29]$$

where the L_{DD} is the distance per jump, so that J_{DD} becomes:

$$J_{DD} = - k_{DD} \, \mathsf{L}_{DD} \, \frac{dn_{DD}}{dx} = - k_{DD} \, \mathsf{L}^2{}_{DD} \, \frac{dC_D}{dx} = - D_{DD} \, \frac{dC_D}{dx} \qquad [2.30]$$

For the type of jumps from H to D one may proceed in a similar way.

$$J_{HD} = k_{HD} \, [n_{HD} - n'_{HD}] \qquad\qquad [2.31]$$

$$J_{HD} = - D_{HD} \, \frac{dC_H}{dx} \qquad\qquad [2.32]$$

For the type of jumps from D to H, Barrer includes the possibility of finding a hole adjacent to the dissolved molecule. The possibility is $(1-\phi)$ where $\phi = C_H/C'_H$; the fractional coverage of the Langmuir site.

$$J_{DH} = k_{DH} \, (n_{DH} \, [1 - \phi] - n'_{DH} \, [1 - \phi]) \qquad\qquad [2.33]$$

similarly:

$$J_{DH} = - D_{DH} \, [\, dC_D \, / \, dx \, (1 - \phi) + C_D \, / \, C'_H \cdot dC_H \, / \, dx \,] \qquad [2.34]$$

The type of jumps from H to H requires that the adjacent hole has to be empty too. It is also necessary that two holes be adjacent and are so placed that jumps between them occur from left to right or from right to left directly (Barrer, 1984). Then, the net flow is:

$$J_{HH} = k_{HH} \, (n_{HH} \, [1 - \phi] - n'_{HH} \, [1 - \phi]) \qquad\qquad [2.35]$$

Similarly:

$$J_{HH} = - D_{HH} \, \frac{dC_H}{dx} \qquad\qquad [2.36]$$

From this treatment Barrer derived the steady state flux as:

$$J = -\,(D_{DD} + D_{DH}\,[1 - \phi]\,) \, \frac{dC_D}{dx} - [D_{HH} + D_{HD} + D_{DH}\,\,(C_D\,/\,C'_H\,)\,] \, \frac{dC_H}{dx} \qquad [2.37]$$

This work has considered the diffusion mechanism on a molecular scale; it will be useful as a guide in interpretation of transient binary gas diffusion, as will be apparent a little later in this chapter.

2.4 TIME LAG METHOD

To investigate a gas diffusion rate under the influence of a second co- or counterdiffusing component, the time lag method may be used. This method was first proposed by Daynes (1920), and refined by Barrer (1939). It allows a monitoring of diffusion from "time zero" until the diffusion process reaches a constant permeation rate. Under most measurement conditions, the subject membrane is initially free of gas. Gas is then introduced on the upstream side of the membrane and the accumulated amount of transported species is measured on the downstream side. In this manner, transient diffusion as well as steady state diffusion can be studied.

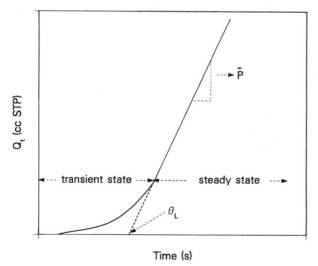

Figure 2.6 Time lag method.

Figure 2.6 is a typical example of a time lag experiment. Accumulated diffusive flux is plotted as a function of time. The initial curve shows the changing diffusion rate in the unsteady state and the linear region shows a constant flux when steady state is reached. Extrapolation of the steady state portion of the figure intersects the time axis; this time is referred to as the time lag and is denoted by Θ (Crank and Park, 1968; Crank, 1975).

For diffusion in a plane sheet, the accumulated amount of diffusing substance Q_t which has passed through the plane sheet in time t is given by:

$$Q_t = \int_0^t J \Big|_{x=l} \cdot dt = \int_0^t - D \left(\partial C / \partial x \right) \Big|_{x=l} \cdot dt \qquad [2.38]$$

where l is the sheet thickness, D is the diffusion coefficient, C is the concentration, and J is local diffusive flux as the gas emerges from unit area of the downstream face, $x = l$, of the membrane. The following relationship for the time lag and diffusion coefficient can be obtained:

$$\Theta = l^2 / 6D \qquad [2.39]$$

where Θ is the time lag. According to Crank (1975), within the accuracy of plotting, the steady state is reached when $t = 0.45 \, (l^2/D)$. Siegel and Coughlin (1972) estimated the relationship between time lag and the time required for a Fick's law diffusion process to reach steady state. They concluded that the proper range of experimental data which should be analyzed for steady state diffusion are those obtained after experimental times larger than 2Θ. These results imply that the time for a specific diffusion process to reach steady state has a fixed relation to the time lag measured.

For a more complicated sorption and diffusion mechanism than simple Fick's law diffusion of a single species, the time lag function may not be so readily obtained, nor will it always be possible to precisely calculate the time to reach steady state. However, experimental time lag can always be obtained from its definition; that is, to extrapolate the accumulated amount (Q_t) from the steady state region until it crosses the time axis. While the numerical ratio between time lag and the time required to reach steady state varies from case to case, for each specific system there should exist a certain fixed ratio for purposes of approximation.

Paul and Koros (1976) developed a time lag model with the partial immobilization assumption. Bhatia and Vieth (1980) developed a slightly more general expression for the time lag where there are dual Langmuirian modes and both species are mobile. From their work it can be concluded that time lag is a function of sorption constants, diffusion coefficients, and gas pressures at the boundary. The precise function varies because of different sorption isotherms and diffusion mechanisms. These efforts demonstrated again that, to measure the time lag, one has only to analyze the steady state part of the diffusion profile in the manner prescribed without knowing details of the transient state diffusion (Frisch, 1957).

For these reasons, the time lag is used as a parameter to indicate the time required for diffusion to reach steady state. But taking the counterdiffusion of two gases as an example, one can only say that, under the same diffusion partial pressure for a given component, if the time lag is larger in an instance, then the diffusion process is considered to reach steady state more slowly; conversely, if the time lag is smaller, the diffusion process is considered to reach steady state more rapidly.

Since the determination of accumulated amount involves an integration of flux, the slope of the linear constant flux region in Figure 2.6 represents the steady state flux. But permeability rather than flux is a more convenient measure, commonly used to characterize steady state diffusion. As already seen, it is defined as flux divided by pressure gradient.

$$\overline{P} = J / [\Delta P / \Delta x] \qquad\qquad [2.40]$$

For the unsteady state diffusion mechanism, no single parameter can be used to describe the process. The instantaneous slope of the experimental curve represents the transient flux. In addition to experimentation, utilization of the instantaneous numerical values generated by computer simulation helps to clarify the transient process, as will be seen presently.

2.5 COUNTERDIFFUSION PERMEABILITY

A general expression for the steady state counterdiffusion permeability can be derived from Fick's law (Jiang, 1987). Consider a diffusion scheme involving, for instance, CO_2 and CH_4; designate A to represent CO_2, and B to represent CH_4. The *position for gas A at x=0* is considered as the *upstream* side and $x = 0$ as the downstream side for the membrane, while *for B, x = 0 is the downstream side* and $x=0$ is the upstream side. Gases A and B counterdiffuse due to partial pressure driving forces, $P_A^0 - P_A^1$ and $P_B^1 - P_B^0$, respectively. At steady state, the flux J_A of component A through a membrane of thickness 0 is given by:

$$J_A = - D_{DA} \partial C_{DA} / \partial x - D_{HA} \partial C_{HA} / \partial x \qquad [2.41]$$

C_{DA} and C_{HA} are the local concentrations of penetrant A in the Henry's law and Langmuir environments, respectively. Since the steady state flux is constant, we have:

$$\int_{0}^{0} J_A dx = - \int_{C_{DA}^1}^{C_{DA}^0} D_{DA} dC_{DA} - \int_{C_{HA}^1}^{C_{HA}^0} D_{HA} dC_{HA} \qquad [2.42]$$

By the assumption that local equilibrium exists at all points in the polymer, C_{HA} can be expressed in terms of C_{DA} and C_{DB} by equating the component partial pressures within the Henry's law and Langmuir environments for gases A and B, respectively. From eqns. [2.13] and [2.14], and [2.15] and [2.16], $P_A = C_{DA} / k_{DA}$ and $P_B = C_{DB} / k_{DB}$. Substituting into C_{HA},

$$C_{HA} = \frac{K_A C_{DA}}{1 + \alpha C_{DA} + \beta C_{DB}} \qquad [2.43]$$

where $K_A = C'_{HA} b_A / k_{DA}$; $\alpha = b_A / k_{DA}$; $\beta = b_B / k_{DB}$. Let $F_A = D_{HA} / D_{DA}$. Substituting for C_{HA} into the integrated form of eqn. [2.42] we have:

$$J_A = \frac{D_{DA}}{0} [C_{DA}^0 - C_{DA}^1 + F_A [(K_A C_{DA}^0 / 1 + \alpha C_{DA}^0 + \beta C_{DB}^0) -$$

$$(K_A C_{DA}^1 / 1 + \alpha C_{DA}^1 + \beta C_{DB}^1)]] \qquad [2.44]$$

For a given gas-polymer system, the permeability $\overline{P}$ is defined as the flux rate divided by the pressure gradient. The steady state permeability of component A can thus be expressed as:

$$\overline{P}_A = \frac{D_{DA}}{P_A^O - P_A^l} \, [C_{DA}^O - C_{DA}^l + F_A \, [\, (K_A \, C_{DA}^O \, / 1 + b_A \, P_A^O + b_B \, P_B^O \,) -$$

$$(\, K_A \, C_{DA}^l \, / 1 + b_A \, P_A^l + b_B \, P_B^l \,) \,]\,] \qquad\qquad [2.45]$$

This equation represents the general expression for permeability in counterdiffusion. When the experiment is performed with a negligible down-stream partial pressure of components A and B, or $P_A^l = 0$, $P_B^O = 0$, eqn. [2.45] can be reduced to:

$$\overline{P}_A = [\, 1 + (\, F_A \, K_A \, / 1 + b_A \, P_A \,) \,] \qquad\qquad [2.46]$$

$$F_B = D_{HB} \, / \, D_{DB}, \quad K_B = C'_{HB} b_B \, / \, k_{DB} \qquad\qquad [2.47]$$

A similar equation can be derived for gas B:

$$\overline{P}_B = D_{DB} \, k_{DB} \, [\, 1 + (F_B \, K_B \, / 1 + b_B \, P_B \,) \,] \qquad\qquad [2.48]$$

Comparing eqns. [2.46] and [2.48] to the permeability of single gas diffusion, eqn. [2.24], one perceives that there is no difference. This is a surprising result of counterdiffusion analysis - with negligible downstream pressure, steady state permeability is exactly the same as for single gas diffusion. The counterdiffusion scheme provides a basis for measuring two permeabilities simultaneously and conveniently.

2.6   FORMULATION OF UNSTEADY STATE BINARY DIFFUSION

Let us go further to consider the transient case. For a two component, dual sorption, dual mobility situation, equations can be written as follows:

$$\partial C_{DA}/\partial t + \partial C_{HA}/\partial t = D_{DA}\partial^2 C_{DA}/\partial x^2 + D_{HA}\partial^2 C_{HA}/\partial x^2 \qquad [2.49a]$$

$$\partial C_{DB}/\partial t + \partial C_{HB}/\partial t = D_{DB}\, \partial^2 C_{DB}/\partial x^2 + D_{HB}\, \partial^2 C_{HB}\, \partial x^2 \qquad [2.49b]$$

These two equations are connected through C_{HA} and C_{HB}, where C_{HA} and C_{HB} are simultaneous functions of C_{DA} and C_{DB}. From the assumption that gases in two different modes are in local equilibrium, P_A and P_B in the Langmuirian isotherm can be replaced by C_{DA} and C_{DB}, as shown in eqn. [2.43]; a similar equation can be written for C_{HB} .

$$C_{HA} = \frac{[C'_{HA}b_A/k_{DA}]\,C_{DA}}{1+\alpha C_{DA}+\beta C_{DB}} \quad ; \quad C_{HB} = \frac{[C'_{HB}b_B/k_{DB}]\,C_{DB}}{1+\alpha C_{DA}+\beta C_{DB}} \qquad [2.50]$$

Assuming only pure gases are present at each side of the membrane, the boundary conditions are simple for either codiffusion or counter-diffusion; to explore a previously unsolved case we chose to focus on the latter.

$$\text{For } t = 0 \qquad \text{All x} \qquad C_A = C_B = 0$$

$$C_{DA} = C_{HA} = C_{DB} = C_{HB} = 0 \qquad [2.51]$$

$$\text{For } t > 0 \qquad \text{At } x = 0$$

$$C_A^0 = C_{DA}^0 + C_{HA}^0$$

$$C_{DA}^0 = k_{DA}\,P_A^0 \qquad C_{HA}^0 = \frac{C'_{HA}\alpha C_{DA}^0}{1+\alpha C_{DA}^0}$$

$$C_B^0 = C_{DB}^0 + C_{HB}^0 = 0$$

$$\text{For } t > 0 \qquad \text{At } x = l$$

$$C_A^l = C_{DA}^l = C_{HA}^l = 0$$

$$C_B^l = C_{DB}^l + C_{HB}^l$$

$$C_{DB}^l = k_{DB}\,P_B^l \qquad C_{HB}^l = \frac{C'_{HB}\,\beta C_{DB}^l}{1 + \beta\,C_{DB}^l}$$

Solution to the above boundary value problem determines the experimentally measured fluxes $J_A(l,t)$, $J_B(0,t)$ through the membrane:

$$J_A [L,t] = - \{D_{DA} \, \partial C_{DA} / \partial x + D_{HA} \partial C_{HA} / \partial x\}_{x=L} \qquad [2.52a]$$

$$J_B [0,t] = \{D_{DB} \, \partial C_{DB} / \partial x + D_{HB} \, \partial C_{HB} / \partial x \}_{x=0} \qquad [2.52b]$$

The accumulated amounts of transported species can be evaluated by:

$$Q_A = \int_0^t J_A [L,t] \, dt \qquad [2.53a]$$

$$Q_B = \int_0^t J_B [0,t] \, dt \qquad [2.53b]$$

A numerical method was used to solve these equations. To simplify the numerical calculation, the following dimensionless groups were introduced:

$\tilde{C}_{DA} = C_{DA} / C_A^0$, $\tilde{C}_{DB} = C_{DB} / C_B^1$, $\tilde{C}_{HA} = C_{HA} / C_A^0$, $\tilde{C}_{HB} = C_{HB} / C_B^1$, $\tilde{C}'_{HA} = C'_{HA} / C_A^0$, $\tilde{C}'_{HB} = C'_{HB} / C_B^1$, $\tilde{b}_A = [b_A / k_{DA}] \, C_A^0$, $\tilde{b}_B = [b_B / k_{DB}] \, C_B^1$, $\tilde{r}_a = D_{HA} / D_{DA}$, $r_b = D_{DB} / D_{DA}$, $\tilde{r}_c = D_{HB} / D_{DA}$, $\mathcal{T} = D_{DA} \, t / L^2$, $X = x / L$.

After substituting the dimensionless variables the differential equations become:

$$\partial \tilde{C}_{DA} / \partial \mathcal{T} + \partial \tilde{C}_{HA} / \partial \mathcal{T} = \partial^2 \tilde{C}_{DA} / \partial X^2 + r_a \, \partial^2 \tilde{C}_{HA} / \partial X^2 \qquad [2.54a]$$

$$\partial \tilde{C}_{DB} / \partial \mathcal{T} + \partial \tilde{C}_{HB} / \partial \mathcal{T} = r_b \, \partial^2 \tilde{C}_{DB} / \partial X^2 + r_c \, \partial^2 \tilde{C}_{HB} / \partial X^2 \qquad [2.54b]$$

and the Langmuirian isotherms become:

$$\tilde{C}_{HA} = \frac{\tilde{C}'_{HA} \, \tilde{b}_A \, \tilde{C}_{DA}}{1 + \tilde{b}_A \, \tilde{C}_{DA} + \tilde{b}_B \, \tilde{C}_{DB}} \qquad [2.55a]$$

$$\tilde{C}_{HB} = \frac{\tilde{C}'_{HB} \, \tilde{b}_B \, \tilde{C}_{DB}}{1 + \tilde{b}_A \, \tilde{C}_{DA} + \tilde{b}_B \, \tilde{C}_{DB}} \qquad [2.55b]$$

For the above boundary value problem, the transient fluxes are defined in dimensionless forms:

$$\mathcal{J}_A [1,\mathcal{T}] = - \{D_{DA} / D_{DA} \, \partial \, \tilde{C}_{DA} / \partial X + D_{HA} / D_{DA} \, \partial \, \tilde{C}_{HA} / \partial X\}_{X=1} \qquad [2.56a]$$

$$\tilde{J}_B\,[0,\mathcal{T}] = \{D_{DB}\,/D_{DA}\,\partial\,\tilde{C}_{DB}\,/\partial X + D_{HB}\,/D_{DA}\,\partial\,\tilde{C}_{HB}\,/\partial X\}_{X=0} \qquad [2.56b]$$

The accumulated amounts of transported species can then be evaluated by:

$$\tilde{Q}_A = \int\,\tilde{J}_A\,[1,\mathcal{T}\,]\,d\mathcal{T} \qquad\qquad [2.57a]$$

$$\tilde{Q}_B = \int\,\tilde{J}_B\,[0,\mathcal{T}\,]\,d\mathcal{T} \qquad\qquad [2.57b]$$

and the counterdiffusion time lags can be evaluated as:

$$\tilde{\theta}_A = \mathcal{T} - \tilde{Q}_A\,/\tilde{J}_A \qquad\qquad [2.58a]$$

$$\tilde{\theta}_B = \mathcal{T} - \tilde{Q}_B\,/\tilde{J}_B \qquad\qquad [2.58b]$$

An analytical solution for the transient problem is not readily obtainable, so the simultaneous partial differential equations were solved by a high accuracy step-wise numerical integration technique. From the basic assumption that constants from pure component measurements can be applied directly to mixed gases, a set of reported constants of carbon dioxide and methane in polycarbonate at 35°C, as listed in Table 2.1, was used for this study (Koros et al., 1977).

TABLE 2.1   DUAL MODE PARAMETERS FOR POLYCARBONATE AT 35°C

GAS	k_D cc gas (STP) cc(Polymer)atm	C'_H cc gas (STP) cc(Polymer)atm	b atm^{-1}	D_D x 10^8cm^2/sec	D_H x10^9cm^2/sec
CO_2	0.6852	18.805	0.2618	6.22	4.85
CH_4	0.1473	8.382	0.0841	1.09	1.258

The transport equations have an associated set of boundary conditions and a set of parameters which include sorption related coefficients (k_{DA}, k_{DB}, b_A, b_B, C'_{HA} and C'_{HB}) and diffusion related constants (D_{DA}, D_{HA}, D_{DB} and D_{HB}). The pressure of carbon dioxide of 10 atm and 10 atm pressure of methane were chosen as reference values. Various pressures of methane were simulated to calculate counterdiffusion with carbon dioxide kept at 10 atm and conversely. Pure carbon dioxide and methane transport were also calculated at corresponding pressures, respectively. The thickness of membrane used in the calculations was $L = 7.62 \times 10^{-3}$ cm and the assumed temperature was 35°C. The equations were then solved by the finite difference method; calculations were carried out on the AS-9000 computer.

2.7 SIMULATION RESULTS

Figures 2.7 and 2.8 show the diffusive fluxes of carbon dioxide and methane, respectively; these flux values are increased when compared to single gas diffusion. The flux also increases with increase of counterdiffusing gas pressure. This explains the shifts in the accumulated flux versus time curves. The flux of carbon dioxide has the more interesting behavior. That is, under counterdiffusion, the flux rises to a maximum value and then slowly decreases to the steady state value (see also Table 2.2). This explains the curvature in the early part of a $\widetilde{Q}_A - \mathcal{T}$ figure, as shown later in Fig. 2.10.

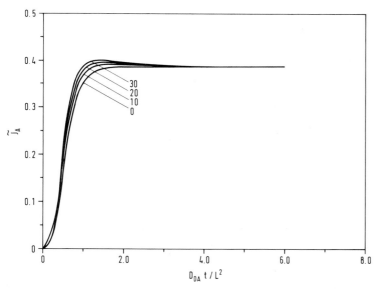

Figure 2.7   Computed flux of carbon dioxide at membrane boundary (CO_2 at 10 atm). Other numbers refer to the pressure of methane (atm).

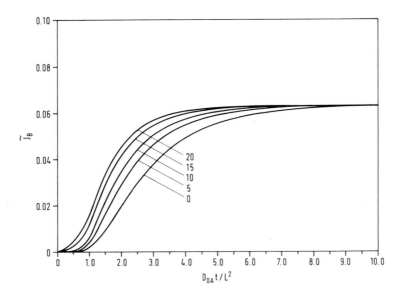

Figure 2.8   Dimensionless flux of methane at membrane boundary ($\times$=0) as a function of dimensionless time.   Methane pressure is 10 atm and other numbers refer to the pressure of carbon dioxide (atm).

TABLE 2.2 COMPARISON OF THE MAXIMUM FLUX AND STEADY STATE FLUX OF
          CARBON   DIOXIDE (A)

P_A (atm)	P_B (atm)	Maximum flux (dimensionless)	Steady state flux (dimensionless)	% increase for maximum flux
10	2.5	0.38807	0.38677	0.35
10	5	0.38945	0.38678	0.69
10	10	0.39202	0.38679	1.35
10	15	0.39426	0.38679	1.94
10	20	0.39628	0.38679	2.46
10	25	0.39813	0.38679	2.94

It is apparent from the above observations that two effects change the component diffusive flux in counterdiffusion. There is a factor which "speeds up" the flux and a factor which "retards" the flux (Koros et al., 1980). For carbon dioxide, the accelerating effect alters the early stage of diffusion and the retarding effect makes its appearance in the later stage. For methane, only the speed up effect (shortened time lags) is obvious from the flux versus time curves; however, the retarding effect does exist and will be discussed later. At steady state these two effects compensate each other and there results a constant flux despite the presence of the counterdiffusing gas. This is consistent with the previous result from the generalization of steady state permeability (eqn. [2.46]).

2.8 EXPERIMENTAL RESULTS

A combination of carbon dioxide at 7.8 atm (100 psig) and methane at 12.22 atm (165 psig) was chosen as the reference run. Carbon dioxide at this pressure counterdiffused against methane at various pressures through poly(ethylene terephthalate) membranes. The methane boundary pressures against carbon dioxide were 4.4 atm, 7.8 atm, 13.2 atm and 18.0 atm. Likewise, methane at 12.22 atm counterdiffused against various carbon dioxide pressures. The counterdiffusing CO_2 pressures against methane were 7.19 atm, 12.22 atm, 15.01 atm, and 18.0 atm. Single component diffusion at the corresponding partial pressure of each mixture component was also performed for purposes of comparison and data estimation.

The accumulated fluxes, Q_B, of methane at 12.22 atm are presented in Figure 2.9. The curves are shifted toward the Q_B axis under the influence of increasing pressure of carbon dioxide. The time lag of pure methane at 12.22 atm is 315 minutes which is reduced to 138 minutes when counterdiffusing with 18.0 atm of carbon dioxide. This dramatically illustrates the effect of the second component that makes the diffusion process reach steady state faster. The linear regions of Figure 2.9 are parallel to each other, demonstrating that the steady state permeability remains constant in spite of the presence of the second component, as predicted by eqn. [2.46] and the limit of the unsteady state simulation.

A similar trend can be observed for carbon dioxide. Figure 2.10 presents the accumulated amounts of carbon dioxide at 7.8 atm. The

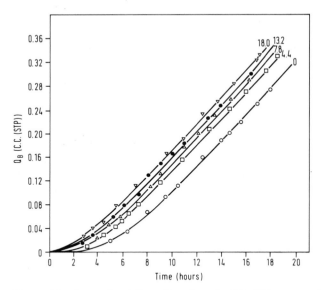

Figure 2.9   Accumulated flux of methane in CO_2/CH_4 counterdiffusion.   Methane at 12.22 atm. Numbers on the curve refer to the pressure of carbon dioxide (atm); zero indicates that there is pure methane diffusion.

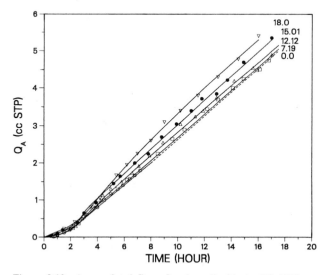

Figure 2.10   Accumulated flux of carbon dioxide in CO_2/CH_4 counterdiffusion.   Carbon dioxide at 7.8 atm.   Numbers on the curve refer to the pressure of methane (atm); zero indicates that there is pure carbon dioxide diffusion.   The dotted line is extended from the steady state region of pure CO_2 diffusion.

curves shift toward the Q_A axis and the linear regions once again parallel each other. In the early part of counterdiffusion, the curves show the characteristic curvature that results from the flux variation, predictable from the simulation work.

The time lags decrease as methane pressure increases. The relative change of these time lags is smaller when compared to CO_2 effects on methane, indicating that carbon dioxide has a stronger accelerating effect on methane, a result of its higher affinity for Langmuir sites; the time lags of methane tend to level off as the carbon dioxide pressure increases. This reveals the possibility of a "saturation effect," in which the effect of the second component appears to reach a limit; i.e., the tendency to provoke an earlier steady state levels out. This, too, will be discussed later.

2.9 THE CHANGE OF LANGMUIRIAN FLUXES

To evaluate the changes in Langmuirian fluxes, consider a differential plane normal to the direction of flow as before. The Langmuir mode flux of gas A in single gas diffusion or counterdiffusion against gas B can be written as follows:
Single gas diffusion:

$$J_{HA} = -\frac{D_{HA}}{\Delta x} [\; C'_{HA} b_A P_A / 1 + b_A P_A \;\Big|_{x+\Delta x} -$$

$$C'_{HA} b_A P_A / 1 + b_A P_A \Big|_x] \qquad [2.59]$$

A and B counterdiffusion:

$$J_{HA} = -\frac{D_{HA}}{\Delta x} [C'_{HA} b_A P_A / 1 + b_A P_A + b_B P_B \Big|_{x+\Delta x} -$$

$$C'_{HA} b_A P_A / 1 + b_A P_A + b_B P_B \Big|_x] \qquad [2.60]$$

Comparing these two equations and assuming the same driving pressure P_A, the difference is the $b_B P_B$ term in the denominator of the isotherm. So, in eqn. [2.60] the flux is not only determined by the concentration of gas A, but is affected by the concentration of gas B as well. The primary influence of B on A is that, in contrast to eqn. [2.59], the concentration term of eqn. [2.60] is decreased by the presence of $b_B P_B$ in the denominator.   The secondary influence is that B has a reversed concentration gradient with respect to A which implies that $b_B P_B \mid_x < b_B P_B \mid_{x+\triangle x}$. Now, a larger value in the denominator makes the concentration term even smaller. The larger value of $b_B P_B$ at position $x+\triangle x$ thus enhances the concentration gradient of A. This increases the A flux in counterdiffusion and speeds up the diffusion process.

The enhancement effect comes in even more importantly in the early transient stage. From Table 2.1, it can be seen that the diffusion coefficient of A is about fivefold the diffusion coefficient of B; the diffusion process of A is therefore much more rapid than that of B. In the beginning of the diffusion process, A migrates in the region near the upstream side of the membrane without much experiencing the presence of B. When A moves into the downstream side of the membrane and has developed its local concentration to a certain level, the concentration of B is relatively undeveloped and has a very steep concentration profile. Therefore, the concentration of B in the denominator of eqn. [2.60] changes substantially from position to position and that further enhances the differences in the concentration gradient of A. The flux of A is thus sped up much more in the transient state and eventually goes to its maximum value.

As gas B moves into the membrane it can speed up gas A as mentioned above, but it can also later retard A because the gradual increase of B concentration locally eventually decreases the concentration of C_{HA} according to the isotherm. Decreasing the concentration of C_{HA} reduces the driving force which eventually causes the flux to decrease. At the same time, the relative difference of $b_B P_B$ from position to position is reduced while B diffuses into the membrane as time goes by, which tends to smooth the previously discussed steep concentration gradient and the "speed-up" effect is reduced. At the steady state these accelerating and retarding influences compensate each other, and the resulting flux remains constant. To complete the understanding of this behavior, the change of flux in the Henry's mode has to be included and will be discussed in the section to follow.

For gas B, the presence of A also tends to increase the concentration gradient governing its flux, thereby increasing the flux rate. The difference between A and B is that, for gas B, with its smaller diffusion coefficient, the process of diffusion is generally sluggish as compared to A. When gas B migrates into the membrane, the concentration level of gas A has already developed to a certain degree and since the A concentration level develops faster, the concentration difference of A from position to position experienced by B is more of a smooth change. Thus, the change of the C_{HB} gradient is influenced by a "relatively constant" local gradient of A concentration. As a consequence, the accelerating and retarding effects operate at about the same level of importance for A on B, either in the early stage of diffusion or in the later stage. The resulting process for B is that it migrates into the membrane, gradually competes for the unrelaxed free volume space which makes its local concentration increase, and its flux eventually rises to its steady state value without going through a maximum transition.

2.10 THE CHANGE OF CONCENTRATIONS AND FLUX COMPENSATION

In this section, we shall move ahead a little to employ some simple results derived in the subsequent section on Effective Diffusivity (see page 41) to consider the steady state concentration profile in both counterdiffusion and single gas diffusion. The concentration profile of gas A in counterdiffusion is given as:

$$[D_{DA}C_{DA} + D_{HA}C_{HA}] = [D_{DA}C_{DA}^O + D_{HA}C_{HA}^O] \, [1 - x/l] \qquad [2.61]$$

while for single gas diffusion:

$$[D_D C_D + D_H C_H] = [D_D C_D^O + D_H C_H^O] \, [1 - x/l] \qquad [2.62]$$

When subject to the appropriate boundary conditions, the right hand sides of the above equations are essentially the same. The presence of the second component B, however, decreases the value of C_{HA} in eqn. [2.61]. In order to maintain the equivalence, the value of C_{DA} in eqn. [2.61] has to be increased. The local concentration level is increased in the Henry's mode, decreased in the Langmuir mode and decreased in total concentration. This type of concentration alternation is also true for the transient state.

The changes in concentration influence the diffusional fluxes. The concentration influence corresponding to the Langmuirian flux change has been discussed in the last section. But, as a result of the concentration change, the Henry's flux is also changed. We already know that the permeabilities of single gas diffusion and counter-diffusion are the same under the same boundary conditions, because the steady state fluxes are equal in each case. The total fluxes in both cases can be split into two parts, as:

$$J_A = J_{DA} + J_{HA} \qquad\qquad\qquad [2.63a]$$

$$= - D_{DA} \partial C_{DA} / \partial x - D_{HA} \partial C_{HA} / \partial x \qquad [2.63b]$$

where J_A is the total flux of A, J_{DA} is the Henrian flux, and J_{HA} is the Langmuirian flux. To maintain a constant total flux, the individual flux J_{DA} is changed to "compensate" for the change of J_{HA}. Since the diffusion coefficients D_{DA} and D_{HA} are constant, the change of concentrations influences the observed flux at the membrane boundary. Thus, it is clear that the observed flux change is due to the concentration factor, not to the mobility factor. The decrease of C_{HA} occurs together with the increase of C_{DA}. Therefore, there is a "compensation" of flux, since, when one mode has a change in flux, the second mode changes its local concentration as well as its gradient to compensate for the reduced local flux in the first mode.

A diffusional flux, of course, is dependent both on the concentration level of a species and its concentration gradient. We have discussed the enhancement of Langmuirian flux, the change of concentrations and the change of Henrian flux. The change of concentration in the steady state is fixed and does not depend on time or position in the membrane; however, in the transient case, the change of fluxes, which depends on the concentration of the second component, does depend on time and position. Consequently, fluxes may instantaneously increase or decrease as functions of these variables.

2.11 EFFECTIVE DIFFUSIVITIES IN MIXED GAS DIFFUSION

The effective diffusivity (D_{eff}) of dual sorption, dual mobility gas transport in glassy polymers was given by Koros, Paul and Rocha (1976). They define the diffusional flux as the product of an effective diffusivity and gradient of total concentration.

$$J = - D_{eff} [C] \frac{\partial C}{\partial x} \qquad\qquad [2.64]$$

For single gas diffusion the partial immobilization model predicts the concentration-dependent effective diffusivity as:

$$D_{eff} = D_D [1 + (KD_H D_D^{-1} / (1 + \alpha C_D)^2)] [1 + (K / (1 + \alpha C_D)^2)]^{-1}$$

$$[2.65]$$

This is mathematically equivalent to the assignment of a concentration-dependent diffusion coefficient. It is a simplified way to view the diffusion process, as one has only to know the total concentration gradient instead of looking into the concentration gradients of both the Henrian and Langmuirian modes individually (see Fig. 2.11).

In a membrane diffusion process, the concentration traverse from the upstream boundary to the downstream boundary (normally at zero) causes a continuously changing profile as a function of position. Since D_{eff} is a function of concentration, it can be evaluated as a function of position in a membrane diffusion process. In what follows, the local D_{eff} will be evaluated involving the boundary condition and the steady concentration profile.   Such a treatment can yield insight into the local physical situation of diffusion in the membrane and can provide a basis to develop D_{eff} for multicomponent transport.

The concluding sections of this chapter require, perforce, a bit of mathematical elaboration which some readers may find at worst,

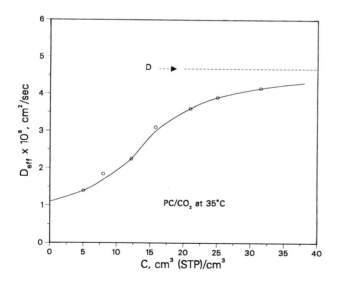

Figure 2.11. Effective diffusivities for carbon dioxide in polycarbonate (at 35°C) (Koros et al., 1976).

intimidating, or, at best, tedious. If so, the author recommends that the reader go directly to Chapter 3. Otherwise, the author is confident that exercise of a little extra patience and effort will be rewarded, by a little deeper insight.

EFFECTIVE DIFFUSIVITY AS A FUNCTION OF TOTAL CONCENTRATION

In previous work (Koros, Paul and Rocha, 1976), the effective diffusivity is expressed as a function of C_D, then plotted as a function of total concentration at the upstream boundary (Fig. 2.11). The explicit form of D_{eff} as a function of total concentration has not yet been presented here. It can be derived from eqn. [2.65], by expressing C_D as a function of total concentration (C). From equation [2.6] and substituting $K=C'_H b/k_D$, $P=C_D/k_D$, and $\alpha = b/k_D$,

$$C = C_D + \frac{KC_D}{1 + \alpha C_D} \qquad [2.66]$$

Therefore:

$$C_D = \frac{-(1+K-C\alpha) \pm [(1+K-C\alpha)^2 + 4\alpha C]^{1/2}}{2\alpha}, \text{ taking the positive root. } [2.67]$$

Replacing the C_D term in eqn. [2.65] by eqn. [2.67], D_{eff} readily becomes a function of total concentration.

EFFECTIVE DIFFUSIVITY AS A FUNCTION OF POSITION

To develop D_{eff} as a function of position the unsteady equation of single gas diffusion, following eqn. [2.49a], is written as:

$$\frac{\partial C_D}{\partial t} + \frac{\partial C_H}{\partial t} = D_D \frac{\partial^2 C_D}{\partial x^2} + D_H \frac{\partial^2 C_H}{\partial x^2} \qquad [2.68]$$

At steady state, the concentration profile can be found as follows:

$$0 = D_D \partial^2 C_D / \partial x^2 + D_H \partial^2 C_H / \partial x^2 \qquad [2.69]$$

Therefore:

$$k_1 = D_D \partial C_D / \partial x + D_H \partial C_H / \partial x \qquad [2.70]$$

$$k_1 x + k_2 = D_D C_D + D_H C_H$$

<u>B.C.</u>

$x = 0$ $k_2 = D_D C^0{}_D + D_H C^0{}_H$

$x = l$ $k_1 l + k_2 = 0$

$$k_1 = -\frac{k_2}{l} = -\frac{1}{l} (D_D C^0{}_D + D_H C^0{}_H) \qquad\qquad [2.71]$$

$$D_D C_D + D_H C_H = -\frac{1}{l} [D_D C^0{}_D + D_H C^0{}_H] x + D_D C^0{}_D + D_H C^0{}_H$$

$$D_D C_D + D_H C_H = [D_D C^0{}_D + D_H C^0{}_H] [1 - x/l]$$

$$C_D + F C_H \equiv \lambda [1 - x/l] \qquad\qquad [2.72]$$

where $C^0{}_D$ and $C^0{}_H$ are the values at the boundary. Expressing C_H in terms of C_D, C_D can be expressed as a function of x (distance along membrane) and the effective diffusivity can subsequently be expressed as a function of x.

Equation [2.72] shows that, at steady state, the summation of Henrian concentration and Langmuirian concentration, each times its diffusion coefficient, is a constant. The total concentration C, which is C_D plus C_H, is not linear in the membrane. If C_H is expressed in terms of C_D, eqn. [2.72] becomes a quadratic function of C_D.

$$C_D^2 + \frac{[1 + FK - \lambda \alpha (1-x/l)]}{\alpha} C_D - \frac{\lambda [1-x/l]}{\alpha} = 0 \qquad\qquad [2.73]$$

$$C_D^2 + k_3 C_D + k_4 = 0$$

$$C_D = [-k_3 \pm (k_3^2 - 4 k_4)] / 2 \quad , \text{ taking the positive root.} \qquad [2.74]$$

Substituting eqn. [2.74] into eqn. [2.65], C_D can be expressed as a function of x and the effective diffusivity can then be expressed as a function of x.

CORRELATION BETWEEN THE LOCAL CONCENTRATIONS OF TWO PENETRANTS

To develop the effective diffusivity model for mixed gases, a procedure similar to the previous one can be followed. The models developed are with reference to component A; nonetheless, this is a general format for both gases A and B. Solving equations [2.49a, 2.49b] for steady state concentration profiles:

$$0 = D_{DA}\partial^2 C_{DA}/\partial x^2 + D_{HA}\partial^2 C_{HA}/\partial x^2 \qquad [2.75a]$$

$$0 = D_{DB}\partial^2 C_{DB}/\partial x^2 + D_{HB}\partial^2 C_{HB}/\partial x^2 \qquad [2.75b]$$

Solving the above equations with boundary conditions:

$$C_{DA} + F_A C_{HA} = [C_{DA}^0 + F_A C_{HA}^0][1 - x/L] \qquad [2.76a]$$

$$C_{DB} + F_B C_{HB} = [C_{DB}^1 + F_B C_{HB}^1][x/L] \qquad [2.76b]$$

let: $[C_{DA}^0 + F_A C_{HA}^0] = \lambda_a$

$[C_{DB}^1 + F_B C_{HB}^1] = \lambda_b$

$\lambda_{ab} = \lambda_a/\lambda_b$

where C^0_{DA}, C^0_{HA}, C^L_{DB} and C^L_{HB} are the concentration values at x=0 and x=L, respectively. Expressing C_{HA} and C_{HB} in terms of C_{DA} and C_{DB}, a quadratic equation, eqn. [2.77], can be obtained to relate C_{DA} and C_{DB}. To obtain this expression, eqn. [2.76b] is substituted into eqn. [2.76a] by equating the position variable x / L. The resulting expression is a local relation between C_{DB} and C_{DA}. It is given as:

$$C_{DB}^2 + \frac{\lambda_{ab} - \lambda_a\beta + \lambda_{ab}F_B K_B + \lambda_{ab}\alpha C_{DA} + \beta C_{DA}}{\lambda_{ab}\beta} C_{DB} +$$

$$\frac{-\lambda_a + (1 + F_A K_A - \lambda_a\alpha)C_{DA} + \alpha C_{DA}^2}{\lambda_{ab}\beta} = 0$$

$$\therefore C_{DB}^2 + C_1 C_{DB} + C_2 = 0 \qquad [2.77]$$

$$C_{DB} = \frac{-C_1 \pm [C_1^2 - 4C_2]^{1/2}}{2} \quad , \text{ taking the positive root.} \quad [2.78]$$

$\partial C_{DB}/\partial C_{DA}$ can then be obtained from straightforward, though tedious, operations on eqn. [2.78]. It is unnecessary to burden the reader with further details of the calculation at this point.

In other cases, the boundary conditions of eqn. [2.75] can be modified to obtain the corresponding correlation of the diffusing gases. For example, in the case of a cocurrent diffusion process which has boundary conditions such that gases A and B are present at the upstream side (x=0) and both have zero concentration at the downstream side (x=l), an equation similar to eqn. [2.76] which comes from eqn. [2.75] can be obtained as:

$$D_{DA}C_{DA} + D_{HA}C_{HA} = [D_{DA}C_{DA}^{0} + D_{HA}C_{HA}^{0}] [1 - x/l] \quad [2.79a]$$

$$D_{DB}C_{DB} + D_{HB}C_{HB} = [D_{DB}C_{DB}^{0} + D_{HB}C_{HB}^{0}] [1 - x/l] \quad [2.79b]$$

From this expression the correlation between the concentrations of two gases in cocurrent diffusion can be obtained by the same procedure.

EFFECTIVE DIFFUSIVITIES IN MIXED GAS COUNTERDIFFUSION

One can define the effective diffusivity of a component (A) in a mixed gas system based on total flux of gas A at steady state:

$$J_A = - D_{eff\,A} [\partial C_A / \partial x] \quad [2.80]$$

This equation can also be written as:

$$J_A = - D_{DA} \,\partial C_{DA}/\partial x - D_{HA} \,\partial C_{HA}/\partial x$$

$$= - D_{DA} [1 + F_A \partial C_{HA}/\partial C_{DA}] [\partial C_{DA}/\partial C_A] [\partial C_A/\partial x) \quad [2.81]$$

$$\therefore D_{eff\,A} = D_{DA} [1 + F_A \partial C_{HA}/\partial C_{DA}] [\partial C_{DA}/\partial C_A] \quad [2.82]$$

Similarly, the effective diffusivity for gas B is:

$$J_B = - D_{eff\,B} [\partial C_B/\partial x] \quad [2.83]$$

$$\therefore D_{eff\,B} = D_{DB} [1 + F_B \,\partial C_{HB}/\partial C_{DB}] [\partial C_{DB}/\partial C_B] \quad [2.84]$$

To solve eqn. [2.82], the $\partial C_{HA}/\partial C_{DA}$ and $\partial C_{DA}/\partial C_A$ terms can be found from sorption isotherm relations:

$$\partial C_{HA}/\partial C_{DA} = \frac{[1+\beta C_{DB}]K_A - K_A C_{DA}\beta\partial C_{DB}/\partial C_{DA}}{[1+\alpha C_{DA}+\beta C_{DB}]^2} \qquad [2.85]$$

From the total concentration of A, $C_A = C_{DA} + C_{HA}$:

$$\partial C_{DA}/\partial C_A = 1/[1+(1+\beta C_{DB})K_A - K_A C_{DA}\beta\partial C_{DB}/\partial C_{DA}]/$$

$$(1+\alpha C_{DA}+\beta C_{DB})^2] \qquad [2.86]$$

Substituting eqns. [2.85] and [2.86] into eqn. [2.82] yields:

$$\frac{D_{eff\,A}}{D_{DA}} = \frac{1+F_A\dfrac{(1+\beta C_{DB})K_A - K_A C_{DA}\beta\partial C_{DB}/\partial C_{DA}}{(1+\alpha C_{DA}+\beta C_{DB})^2}}{1+\dfrac{(1+\beta C_{DB})K_A - K_A C_{DA}\beta\partial C_{DB}/\partial C_{DA}}{(1+\alpha C_{DA}+\beta C_{DB})^2}} \qquad [2.87]$$

This is the general effective diffusivity equation for mixed gas diffusion.   At the boundary the equation simplifies:

At $x = 0$, $C_{DB} = 0$, and, $\partial C_{DB}/\partial C_{DA} \neq 0$

$$\frac{D_{eff\,A}}{D_{DA}} = \frac{1+F_A\dfrac{K_A - K_A C_{DA}\beta\partial C_{DB}/\partial C_{DA}}{(1+\alpha C_{DA})^2}}{1+\dfrac{K_A - K_A C_{DA}\beta\partial C_{DB}/\partial C_{DA}}{(1+\alpha C_{DA})^2}} \qquad [2.88]$$

At $x = l$, $C_{DA} = 0$

$$\frac{D_{eff\,A}}{D_{DA}} = \frac{1 + F_A \dfrac{(1 + \beta\,C_{DB})\,K_A}{(1 + \beta\,C_{DB})^2}}{1 + \dfrac{(1 + \beta\,C_{DB})\,K_A}{(1 + \beta\,C_{DB})^2}} \qquad [2.89]$$

which is the result previously obtained (Koros, Paul and Rocha, 1976).

DISCUSSION OF EFFECTIVE DIFFUSIVITIES IN MIXED GAS COUNTERDIFFUSION

In eqn. [2.87], $\partial C_{DB}/\partial C_{DA}$ can be analytically determined from an equation which contains the boundary conditions and is a function of position. To present this model, $D_{eff\,A}$ is plotted as a surface in a three-dimensional plot; one coordinate displays the pressure of the second gas and another coordinate, the position in the membrane. A computer program was used to calculate effective diffusivities. At given boundary pressures, P_A and P_B, D_{eff} was calculated by assigning a specific position in the membrane. At each position local values of C_{DA} and C_{DB} were calculated by solving eqns. [2.76a] and [2.76b] simultaneously and these local concentrations were then substituted to calculate the local effective diffusivities. By using the parameters listed in Table 2.1, the effective diffusivities of carbon dioxide and methane were computed. Figure 2.12 shows D_{effB} for CH_4 at 10 atm with various CO_2 pressures and positions in the membrane. Note that in this figure the position at $x=0$ is the upstream side for the reference gas and is the downstream side for the gas whose pressure is varied.

The effective diffusivity changes with respect to the pressure of the second component and position. For counterdiffusion, the membrane microstructure can be viewed in the same manner as a multilayer material. For one gas the presence of a second gas alternates the unrelaxed free volume distribution of each layer causing the effective diffusivity changes with respect to the pressure of the second component and position. In Figure 2.12 the local D_{eff} of one component at 10 atm in a two component mixture is generally decreased in the upstream side and increased in the downstream side when the pressure of the second component is increased. Two trends exist inside the membrane: the change of D_{eff} with respect to the pressure of the second component and with position. The geometric object revealing these trends (Fig. 2.12) illustrates something of the artistic content residing in the structure of the model.

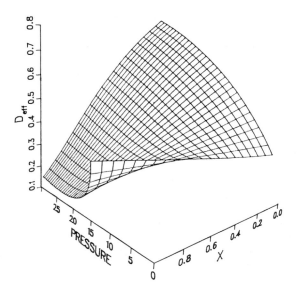

Figure 2.12. Effective diffusivity surface of methane (at 10 atm). Axes are: DEF - effective diffusivity of methane,   x10^8 cm^2/sec; Pressure - carbon dioxide pressure, atm;   X - membrane cross-section (dimensionless);   BC - pure CH$_4$ at X=0, pure CO$_2$ at X=1.

With these tools now in hand, it is time to move on to a consideration of applications of dual sorption theory. This follows directly in the next chapter which includes a consideration of membrane-penetrant structural interactions in reverse osmosis transport.

2.12 NOMENCLATURE

b	hole affinity constant, atm^{-1}
C	concentration, cc gas(STP)/cc polymer
C_D	concentration by normal dissolution, cc gas(STP)/cc polymer
C_H	concentration by hole-filling, cc gas(STP)/cc polymer
C'_H	hole saturation constant, cc gas(STP)/cc polymer
D	diffusion coefficient for the free species, cm^2/sec
D_D	diffusion coefficient of gas in the Henry's law environment, cm^2/sec
D_{eff}	effective diffusivity, cm^2/sec
D_H	diffusion coefficient of gas in the Langmuirian environment, cm^2/sec

D_{T1}	thermodynamic diffusion coefficient of normally dissolved penetrant, cm^2/sec
D_{T2}	thermodynamic diffusion coefficient of adsorbed penetrant, cm^2/sec
F	$= D_H/D_D$, dimensionless
$\triangle H_D$	enthalpy of dissolution, kcal/g mole
$\triangle H_S$	enthalpy of sorption, kcal/g mole
J	penetrant flux, cc(STP)/sec cm^2
J_D	flux in the Henry's law environment, cc(STP)/sec cm^2
J_H	flux in the Langmuirian environment, cc(STP)/sec cm^2
K	$= C'_H \, b/k_D$, dimensionless
k_D	Henry's law solubility constant, cc(STP)/cc polymer atm
k_{D_0}	constant defined in eqn. [2.12], cc(STP)/cc polymer atm
k_1	forward rate constant, cm^2/cc(STP) time
k_{-1}	backward rate constant, $time^{-1}$
$\mathcal{l}$, or occasionally, L,	membrane thickness parameter, cm
n_0	local concentration of immobilizing sites at time t = 0
n_t	local concentration of immobilizing sites at time t
N_0	total immobilizing sites
p	pressure, atm
p_f	final pressure, atm
p_0	initial pressure, atm
$\underline{p_t}$	pressure at time t, atm
P	permeability coefficient, cc(STP)/cm sec atm
	The above notation represents the system for a single component. For a binary system, the subscripts A and B are attached. For example: C_A is the total solubility of CO_2; C_B is total solubility of CH_4; P_A is the partial pressure of A, etc.
Q	accumulated amount, cc(STP)
R	gas constant, kcal/g mole °K
$\mathcal{T}$	dimensionless time; e.g., $D_{DA}t/L^2$
T	temperature, °K
t	time, sec
x	distance from the upstream side of the membrane, cm
α	$= b/k_D$, (single component diffusion)
α	$= b_A/k_{DA}$ (counterdiffusion)
β	$= b_B/k_{D\,B}$
θ	dimensionless time parameter, Dt/L^2
θ'	modified dimensionless time parameter
μ	chemical potential of penetrant
Φ	dimensionless pressure decay $(p_0-p_t)/(p_0-p_f)$

2.13 REFERENCES

Assink, R.A., *J. Polym. Sci.*, Polym. Phys. Ed. **13**, 1665 (1975).

Barrer, R.M., *Trans. Farad. Soc.*, **35**, 628 (1939).

Barrer, R.M., *J. Mem. Sci.*, **18**, 25 (1984).

Barrer, R.M., J.A. Barrie and J. Slater, *J. Polym. Sci.*, **27**, 177 (1958).

Barrie, J.A., A.S. Michaels and W.R. Vieth, *J. Polym. Sci.*, **13**, 859 (1975).

Bhatia, D. and W.R. Vieth, *J. Mem. Sci.*, **6**, 351 (1980).

Chern, R.T., W.J. Koros, E.S. Sanders, S.H. Chen and H.B. Hopfenberg, *Industrial Gas Separation*, ACS Symposium Series 223, Am. Chem. Soc., Washington, D.C. (1983).

Chern, R.T., W.J. Koros, E.S. Sanders and R. Yui, *J. Mem. Sci.*, **10**, 219 (1982).

Crank, J., *The Mathematics of Diffusion*, Oxford University Press: London (1956).

Crank, J., *The Mathematics of Diffusion*, 2nd ed., Clarendon Press: Oxford (1975).

Crank, J. and G.S. Park, *Diffusion in Polymers,* Academic Press: New York (1968).

Daynes, H.A., *Proc. Roy. Soc. Lond.*, Ser.A **97**, 286 (1920).

Eilenberg, J.A. and W.R. Vieth, *J. Appl. Polym. Sci.*, **16**, 945 (1972).

Fenelon, P.J., *Polym. Eng. and Sci.*, **13**, 440 (1973).

Frederickson, G.H. and E. Helfand, *Macromolecules*, **18**, 2201 (1985).

Frisch, H.L., *J. Phys. Chem.*, **61**, 93 (1957).

Frisch, H.L. and S.A. Stern, *CRC Critical Reviews in Solid State and Material Science*, vol. 11, issue 2, 123 (1983).

Jiang, Y.S., Ph.D. Thesis in Chemical and Biochemical Engineering, Rutgers U. (1987).

Komiyama, J. and T. Iijima, *J. Polym. Sci.*, Part A-2, **12**, 1465 (1974).

Koros, W.J., *J. Polym. Sci.*, Polym. Phys. Ed. **18**, 981 (1980).

Koros, W.J., A.H. Chan and D.R. Paul, *J. Mem. Sci.*, **2**, 165 (1977).

Koros, W.J., R.T. Chern, V.T. Stannett and H.B. Hopfenberg, *J. Polym. Sci.*, **20**, 300 (1980).

Koros, W.J. and D.R. Paul, *J. Polym. Sci.*, Polym. Phys. Ed. **16**, 1947 (1978).

Koros, W.J. and D.R. Paul, *J. Polym. Sci.*, Polym. Phys. Ed. **16**, 2171 (1978).

Koros, W.J., D.R. Paul and A.A. Rocha, *J. Polym. Sci.*, Polym. Phys. Ed. **14**, 687 (1976).

Koros, W.J., G.N. Smith and V.T. Stannett, *J. Appl. Polym. Sci.,* **26**, 159 (1981).

McCandless, F.P., *Ind. Eng. Chem. Proc. Des. Dev.,* **11**, 470 (1972).

Meares, P., *J. Am. Chem. Soc.,* **76**, 3415 (1954).

Michaels, A.S., W.R. Vieth and J.A. Barrie, *J. Appl. Phys.,* **34**, 1 (1963).

Michaels, A.S., W.R. Vieth and J.A. Barrie, *J. Appl. Phys.,* **34**, 13 (1963).

Paul, D.R., *J. Polym. Sci.,* Part A-2 **7**, 1811 (1969).

Paul, D.R. and W.J. Koros, *J. Polym. Sci.,* Polym. Phys. Ed. **14**, 675 (1976).

Petropoulos, J.H., *J. Polym. Sci.,* Part A-2 **8**, 1797 (1970).

Pye, D.G., H.H. Hoehnn and M. Panner, *J. Appl. Polym. Sci.,* **20**, 287 (1976).

Pye, D.G., H.H. Hoehnn and M. Panner, *J. Appl. Polym. Sci.,* **20**, 1921 (1976).

Raucher, D. and M.D. Sefcik, *Industrial Gas Separation,* ACS Symposium Series 223, Am. Chem. Soc., Washington, D.C. (1983).

Sanders, E.S., and W.J. Koros, *J. Polym. Sci.,* Polym. Phys. Ed. **24**, 175 (1986).

Sanders, E.S., W.J. Koros, H.B. Hopfenberg and V.T. Stannett, *J. Mem. Sci.,* **13**, 161 (1983).

Sanders, E.S., W.J. Koros, H.B. Hopfenberg and V.T. Stannett, *J. Mem. Sci.,* **18**, 53 (1984).

Sherwood, T.K., R.L. Pigford and C.R. Wilke, *Mass Transfer,* p.5, McGraw Hill (1975).

Siegel, R.D. and R.W. Coughlin, *Recent Advances in Separation Techniques,* AIChE Symposium Series, **68**, 58 (1972).

Tshudy, J.A. and C. von Frankenberg, *J. Polym. Sci.,* Part A-2 **11**, 2027 (1973).

Vieth, W.R., "A Study of Poly(ethylene terephthalate) by Gas Permeation," Sc.D. Thesis, MIT, Cambridge, Massachusetts (1961).

Vieth, W.R., C.S. Frangoulis and J.A. Rionda, Jr., *J. Colloid Interface Sci.,* **22**, 454 (1966).

Vieth, W.R., J.M. Howell and J.H. Hsieh, *J. Mem. Sci.,* **1**, 177 (1976).

Vieth, W.R. and K.J. Sladek, *J. Coll. Sci.,* **20**, 1014 (1965).

Ich denke, was ich will und
    was mich beglücket
Doch alles in der Still und
    wie es sich schicket
Mein Wunsch und Begehren
    kann niemand verwehren,
es bleibet dabei :
    die Gedanken sind frei!

Trad. Ballad

3

APPLICATIONS OF MEMBRANE DIFFUSION-REACTION PRINCIPLES

3.0 INTRODUCTION

At this juncture, it seems worthwhile to set down a few more definitions. According to Webster's New World Dictionary, the source for definitions appearing earlier in this book, a *theory* is " a formulation of underlying principles of certain observed phenomena which has been verified to some degree" while a *model* is "a style or design representing an existing or planned object."

When the author refers to dual sorption theory, he is subscribing to the former definition. That this abstraction is a valuable starting point for analysis of phenomena beyond those encountered with penetrant transport in glassy polymers where the *dual sorption model* is still evolving is clear. That it is of special value in analyzing such important practical processes as reverse osmosis, monomer migraton, enzyme immobilization, as well as in controlled release of drugs, furnishes added interest (Vieth et al., 1976).

3.1 NONEQUILIBRIUM MOLECULAR CHARACTERISTICS OF GLASSY POLYMERS

From the discussions of the previous chapter, it is clear that the inherent nonequilibrium structures of glassy polymers and their effects on polymer properties are of lively interest to scientific researchers. The resulting slow molecular relaxation of chain segments towards an equilibrium state (physical aging) has been shown to affect various important macroscopic properties and practical applications of this class of materials. Many such applications involve sorption and transport of small molecules as fundamental processes. These processes, on the other hand, can be measured and analyzed to transparently reflect the microstructure of polymeric glasses under metastable conditions. Time-dependent sorption and transport properties of small molecules can, therefore, be employed as exquisite probing tools in the quest for a better understanding of the molecular nature of polymers, in general.

In particular, the transient structure and property changes that occur in glassy polymers during annealing or physical aging below the glass-transition temperature, T_g, are of much current research interest. This time-dependent behavior, also referred to as volume or enthalpy relaxation (or recovery), originates from the nonequilibrium nature of the glassy state.   From an applications point of view, it is apparent that this time-dependent phenomenon is the controlling factor in predicting the overall behavior as well as the practical performance of glassy polymeric material as time elapses.   Current research in this area has been focusing on three interrelated aspects: i)   spectroscopic studies of the molecular states in polymer glasses;   ii)   the development of a molecular theory capable of describing as well as predicting the aging behavior, and;   iii)   macroscopic property changes due to structural relaxation and their relationships to the microscopic molecular states.

The nonequilibrium behavior or physical aging of glassy polymers has been shown to originate from the kinetic nature of the glass transition (Tant and Wilkes, 1981).   As a polymer specimen is cooled through the rubber-glass transition from the melt, molecular mobility is rapidly reduced, as evidenced by the sudden increase in

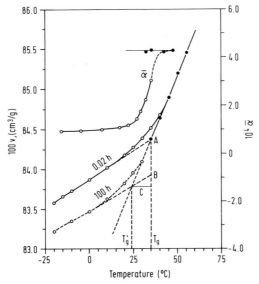

Figure 3.1   Temperature dependence of the specific volume of PVAc in the glass transition range (Kovacs, 1958).

various associated macroscopic properties such as viscosity and modulus.   If the specific volume and enthalpy are simultaneously monitored during this cooling process, the thermal expansion coefficient as well as the heat capacity of the sample are observed to be quickly reduced following the transition region.   This is illustrated in Figure 3.1 (Kovacs, 1958).   In fact, the molecules become *frozen* into a nonequilibrium state as temperature decreases below the glass transition temperature regardless of any practicably retarded cooling rate.

The change in segmental mobility of the polymer chains upon traversing T_g can be illustrated from transport properties as well (Meares, 1954).   In Figure 3.2, the Arrhenius plots of the diffusion coefficients for argon and oxygen in poly(vinylacetate) (PVAc) in a temperature range encompassing the glass transition are shown to be composed of three distinct regions of differing slopes, indicating different activation energies in the three temperature regions.   Meares interpreted these results to mean that, at the upper transition temperature, which coincides with T_g in the Arrhenius plot, the polymer microstructure changes partially from a *liquid-like* phase to an *amorphous solid* or *glass-like* phase. The *dimensions* of these liquid-like regions are reduced with temperature until a critical size is

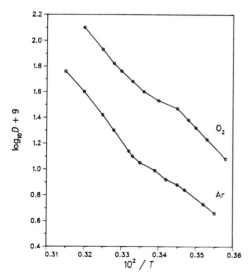

Figure 3.2   Arrhenius plots for the diffusivities of oxygen and argon in a temperature range encompassing the glass transition (Meares, 1954).

reached below which rotation is completely forbidden.   The lower transition temperature essentially characterizes the complete cessation of segmental rotation in these liquid-like regions.

Although the mobility of individual chain molecules is drastically reduced in comparison to that in the rubbery state, it remains finite as a function of temperature and pressure.   Thus, there always exists a working thermodynamic potential or driving force to further decrease the free energy of the system.   Indeed, state functions such as enthalpy, entropy and volume have been found to decrease at different rates with annealing (or aging) time under various isothermal and isobaric conditions at temperatures below T_g (Struik, 1978). Parallel to these changes are the changes in macroscopic physical and mechanical properties such as the rise in density, yield stresses, elastic modulus, and the drop in impact strength, fracture energy, ultimate elongation and creep rate (Petrie, 1975).   In other words, a transition from ductile to brittle behavior is observed.   At the molecular level, spectroscopic studies have indicated that the concentration of the *lower energy conformers* increases as a function of sub-T_g annealing time and temperature (Ito et al., 1978; Moore et al., 1981).   Moreover, a number of experimenters have observed that sorption and transport characteristics of small gas molecules in glassy polymers are also influenced by the nonequilibrium nature of this class of materials (Chen, 1974, 1981; Chan and Paul, 1980a, b; Toi et al., 1985).   Since the mechanisms of sorption and transport of small penetrants are fundamentally related to the molecular nature of polymers in general, it is clear that these processes can be appropriately utilized to probe the nonequilibrium molecular nature of glassy polymers.

3.2 TRANSPORT EFFECTS

The characteristic quasi-equilibrium sorption behavior of gases and vapors in a metastable glassy polymer is shown again in Figure 3.3. Here, it is observed that the total concentration of sorbed carbon dioxide in a glassy polymer, such as poly(ethylene terephthalate) (PET), versus penetrant pressure is concave to the pressure axis for a wide range of temperature below T_g (Koros and Paul, 1978).   (T_g of the samples used to gather these data is about 85°C.)

Hence, unlike the simple sorption behavior of small gas molecules in rubbery polymers at low penetrant pressures or activities

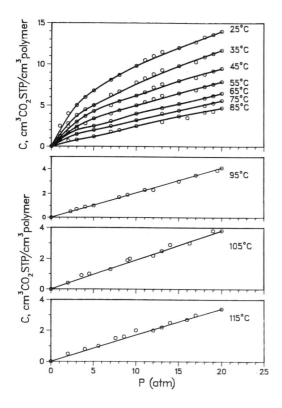

Figure 3.3   Carbon dioxide sorption isotherms in PET between 25 and 115°C (Koros and Paul, 1978).

(which can be adequately quantified by Henry's law in many cases), it is necessary to include an additional Langmuir-type term, coupled with the Henrian term, in order to accurately describe the overall equilibrium sorption characteristics in glassy polymers (e.g., Michaels et al., 1963).

Eilenberg and Vieth (1972) suggested that if one were to assume that CO_2 exists at roughly its liquid density in polymer microvoids, then the Langmuir capacity constant would provide a means of measuring unrelaxed volume.   Koros and Paul (1978) successfully related the Langmuir capacity term to the excess volume of the glassy state as expressed by:

$$C'_H = \frac{[V_g - V_1]\, \rho^*}{V_g} \qquad [3.1]$$

where ρ^* is the "liquid-like" molar density of CO_2 at the point of complete saturation of the Langmuir capacity of the polymer, V_g is the specific volume of the glass and V_1 is that extrapolated from the equilibrium *liquidus* line at the same temperature. ($[V_g-V_1]$ / V_g is, therefore, the unrelaxed or excess free volume fraction, as shown in Fig. 3.4.) An equivalent expression was also introduced as:

$$C'_H = \rho^* \, [\, dV_1 / dT - dV_g / dT \,] \, [\, T_g - T] / V_g \qquad [3.2]$$

According to these authors, the association between the existence of the excess frozen volume below T_g and the presence of the Langmuir term is further substantiated by quasi-static sorption data above T_g. Only the linear Henrian behavior is observed above T_g, as shown in Figure 3.3 (Michaels et al., 1963; Koros and Paul, 1978). This experimental evidence sufficiently supports the conclusion that sorption characteristics clearly reflect the microstructure of glassy polymers.

By varying the experimental temperature and the penetrant upstream pressure in steady state and transient permeation experiments, several researchers (Michaels et al., 1963; Koros and Paul, 1978) have drawn further conclusions regarding the microstructure and molecular mobility of glassy polymers in a quasi-equilibrium state from the permeability and the time lag (a measure of the apparent diffusivity). According to these authors, the Langmuir sorption and transport mode fades away as frozen "microvoids" *collapse* and these microscopic subregions become mobile above the glass transition

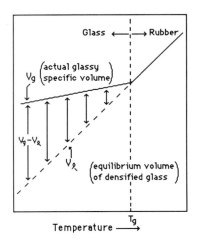

Figure 3.4 Schematic representation of the excess volume (V_g-V_1) of a glassy polymer.

temperature.   In addition, as put forth by Meares   (1954) previously, a distinct discontinuity in the slope of the Arrhenius plot of the Henrian mode diffusivity near T_g might imply that the diffusional jump distance increases above T_g because of an increased zone of activation in the rubbery state relative to the highly restricted, rigid amorphous state.

3.3 HISTORY-DEPENDENT SORPTION AND TRANSPORT BEHAVIOR

Sorption and transport processes are known to be highly dependent on the previous history of a metastable glassy polymer specimen.   In fact, sorption measurements performed on various "as-received" samples at elevated pressures and at temperatures well below T_g often produced very scattered data.   This problem can be resolved by erasing the different previous histories associated with the various specimens by collectively reassigning them a new short-term history. This can be accomplished by "conditioning" or "soaking" the specimens to be studied at a fixed temperature and a high penetrant pressure below which "equilibrium" sorption experiments are to be carried out (Koros and Paul, 1978; Wonder and Paul, 1979).

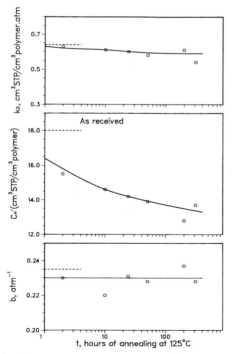

Figure 3.5   Effect of annealing time on dual sorption parameters (Chan and Paul, 1979).

Previous histories which include annealing and conditioning were found to have opposing effects on the subsequent sorption characteristics for a given metastable glassy polymer sample. For the cases of PET and polycarbonate (PC) at 35°C, the Langmuir capacity term C'_H was observed to increase with higher conditioning pressure and lower temperature (Chan and Paul, 1979; Koros and Paul, 1980). The reverse occurs as annealing time is prolonged at lower temperature. The effect of annealing time on C'_H, b and k_D extracted from sorption experiments at 35°C in a CO_2 / PC system is shown in Figure 3.5. The PC sample was annealed at 125°C, about 20°C below T_g. The affinity constant b is unchanged, as expected, since it depends more on the nature of the chemical interaction between polymer and penetrant at a given site. Although these authors subsequently assumed that k_D is invariant with respect to changes in the previous histories in these examples, it appeared that k_D slightly decreased at longer annealing time. The specific volume was shown to decrease with C'_H also. Moreover, as can be seen in Figure 3.6, the relaxation enthalpy increases as C'_H decreases with annealing time, indicating a possible relationship between these two quantities. However, Chan and Paul concluded that there may not be a direct and unique fundamental correlation between the two responses due to annealing.

The changes in permeability and diffusional time lag due to conditioning and annealing were found to depend on the specific nature as well as the present condition of the polymers. In the case of PC (Wonder and Paul, 1979; Chan and Paul, 1980), for instance, sub-T_g

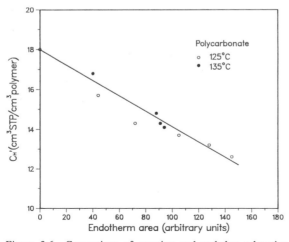

Figure 3.6 Comparison of sorption and enthalpy relaxation (Chan and Paul, 1980a).

annealing lowered both permeability and time lag while conditioning caused the time lag to increase but may also have depressed permeability. While the changes in time lag were significant, a definite conclusion for the permeability results in these studies might be difficult to provide since the level of reproducibility in the data is close to the standard of deviation shown for it. In contrast, preliminary experimental data from our laboratory on CO_2-conditioned PET shows that permeability increases substantially, as indicated in Table 3.1 (Dao, 1987).

Table 3.1 CO_2 Permeability in Mylar® Films @26°C

Film type	Permeability (cc/m^2/day) (atm/mil)
Unconditioned Mylar®	312.6
Conditioned Mylar®	514.3

3.4 MOLECULAR CHARACTERIZATION AND HISTORY-DEPENDENT AND TIME-DEPENDENT MOLECULAR STATES

Since substantial work has already been completed on the molecular characterization of PET as well as its sorption and transport properties, it seems logical to employ this material for case study purposes here. The microstructure of PET is fairly well defined, probably second to poly(ethylene) (PE) in this regard. Unlike a PE chain, however, which is non-polar and has a flexible aliphatic backbone, PET is more polar with a rigid backbone containing aromatic phenyl rings. Using infrared spectroscopy, two distinctive conformers are found to co-exist in completely amorphous as well as semi-crystalline glassy PET samples at ambient conditions (Heffelfinger and Schmidt, 1965; Lin and Koenig, 1982): the *gauche* conformers which constitute a large fraction of the amorphous phase, and the extended *trans* conformers which distribute within the "sea" of the amorphous phase as crystallites and less ordered regions (Heffelfinger and Schmidt, 1965).

The variation in the molecular conformation due to sample conditioning has not yet been experimentally explored. The effects of sample stretching and annealing both above and below T_g have been studied using regular infrared as well as Fourier transform infrared (FTIR) spectroscopy (Heffelfinger and Schmidt, 1965; Lin and Koenig,

1982; Ito et al., 1978; Moore et al., 1981). From the experimental data obtained via X-ray diffraction, infrared, density and stress-strain measurements on PET films, Heffelfinger and Schmidt (1965) explained that, as an initially unoriented cast film is longitudinally stretched in the machine direction (M.D. stretched), the molecules align along the stretch direction, the individual segments becoming extended; i.e., some *gauche* PET is converted into *trans* PET. Crystallization may also occur during the stretching process, but it was observed that more *amorphous trans* conformers are formed by stretching. This amount of *trans*, which is a measure of the "tautness" of the amorphous regions, should also be directly related to mechanical properties of the film.

The opposite molecular behavior occurred in the case of annealed films. Moore et al. (1981) concluded that the fraction of the *gauche* conformation of the glycol moiety increases upon sub-T_g annealing of melt-quenched amorphous PET film at 50°C. An earlier study done by Ito et al. (1978) showed that the *trans* fraction decreased. Annealing at temperatures above T_g also increases the population of *trans* conformers. However, since crystallization is favored within this temperature range in the case of PET, a large fraction of *trans* conformers tends to form crystals (Lin and Koenig, 1982).

3.5 SORPTION AND TRANSPORT BEHAVIOR IN CONDITIONED GLASSY POLYMERS

Far below the glass transition temperature, a glassy polymer behaves as though it is property-invariant with time and its status is thus termed metastable. Near, but still below T_g, many macroscopic property changes are more rapid, quite pronounced and experimentally detectable. Sorption and transport behavior can conveniently be used to monitor the change in the molecular structure and mobility of glassy polymers with time, as exemplified in the following examples.

The effects of aging and conditioning on transport properties of poly(vinylacetate), PVAc, were experimentally studied by Toi et al. (1985) at a fairly low upstream pressure, ca. 10 cm Hg of carbon dioxide. In this work, the apparent permeability and diffusivity of carbon dioxide were continuously measured at various temperatures below T_g (ca. 30°C) after the PVAc film was slowly quenched from the rubbery state. As shown in Figure 3.7, the apparent permeation and diffusion coefficients decrease with physical aging time. These changes may be attributed to reduction in the degree of segmental mobility which is directly related to the level of free volume. Subsequent

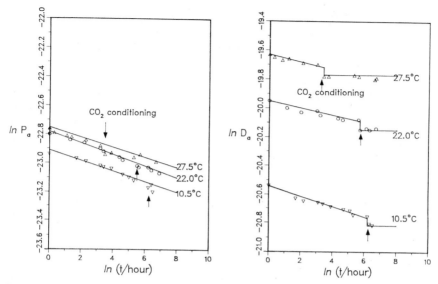

Figure 3.7   Aging effect on the apparent diffusivity and permeability of CO_2 in PVAc at three temperatures.   Arrows show conditioning time (Chan and Paul, 1980a).

exposure to one atmosphere of carbon dioxide substantially lowered the apparent diffusivity but did not alter the trend of the permeability change.   Although the experimental data and technique are quite good, no clear explanations were put forth and no modeling work was attempted.   If the solubility were also to be measured, sufficient data would then be in hand for further modeling efforts.

Earlier works by Chen (1974, 1981) also showed that the diffusivity of methane and propane in PC and PET films decreased with time.   In the later work, the diffusivity of propane at 10°K below the T_g of a polycarbonate sample was shown to decrease linearly with time on a double-logarithmic plot, but it leveled off after some time.   He demonstrated that the higher diffusion rate could be recovered by reheating the sample above T_g and re-cooling it to the same temperature. He concluded that such behavior indicated a reduction in segmental mobility (instead of structural formation as implicated by other authors), which is also responsible for the change in mechanical properties due to physical aging of glassy polymers.   A linear relationship between the diffusivity of methane and the relative impact strength for three polymers (PC, PET, PMMA) was obtained.   Advocating the free-volume theory, he also concluded that the reduction in the specific volume, and in turn the free volume, gave rise to the decrease in the diffusion rate according to the free-volume theory.

As mentioned above, the work of Chan and Paul (1979, 1980) has provided some insights into the effects of sub-T_g annealing on the changes of sorption and transport parameters. Of particular interest were the relations between the change of the Langmuir capacity C'_H and that of the relaxation enthalpy with annealing time. The experimental data plotted in Figure 3.6 suggested a possible mutual, fundamental relationship between these two properties as they vary *during* annealing time.

Instead of raising the experimental temperature to accelerate the aging process as described earlier, other disturbances such as stretching or conditioning may be used to "rejuvenate" the polymer specimen as both actions induce internal stresses. Faster relaxation response can thus be observed. As can be demonstrated in the works of Levita and Smith (1981) and Smith (1983), the rejuvenation and isothermal recovery processes can be experimentally observed by transport properties. They measured the dependence of the true stress, the apparent carbon dioxide permeability $\overline{P}$, diffusivity D, and the ratio $\overline{P}/D$ on tensile strain as it is applied in a stepwise manner on a biaxially oriented poly(styrene) (Trycite) film at a strain rate of 0.04/min at 50°C. The transport coefficients first increase rapidly with the strain, but no further change occurs beyond the yield strain. These increases were ascribed to the increase in the free-volume of the sample with strain. The strain dependence of $\overline{P}$ and D for Ar, Kr, Xe and nitrogen at 50°C was obtained from the initial slopes. Except for carbon dioxide, the values of $(\overline{P} - \overline{P}_0)/\overline{P}_0 \epsilon$ and $(D - D_0)/D_0 \epsilon$ (where $\overline{P}_0$ and D_0 are values at zero strain) showed that they are independent of the molecular diameters. These authors therefore concluded that the size distribution of the free-volume elements in the specimen is not distorted under the influence of small tensile strain.

The apparent gas permeability and diffusivity coefficients in stretched film decrease linearly with time at a constant strain on double-logarithmic plots. The decrease in $\overline{P}$ and D in percent per decade of time are the relaxation rates. Levita and Smith observed that the decay in transport properties was much less than that of the rate of stress, therefore concluding that only volume recovery is responsible for the reduction in transport rates. Regarding the variation in the distribution of the free-volume elements, these authors plausibly argued that the larger free-volume elements decrease at a faster rate than the smaller ones, thus their distribution is skewed towards small elements as time elapses. This conclusion is supported by the fact that the rate of decay in the transport process of xenon is three times greater than that of carbon dioxide due to the larger transverse (or effective) diameter of xenon as compared to that of the linear configuration of carbon dioxide, even though the average molecular diameter of the two species is about the same. Also found was that carbon dioxide molecules at the Langmuir sites were immobile according to the dual mode transport model, and that k_D, C'_H and D_D increased as a result of stretching.

Indeed, it has been substantiated from many previous works that conditioning tends to dilate glassy polymers by increasing the Langmuir sorption capacity. Our preliminary experimental data showed that the steady state permeation rate of a 0.92 mil Mylar® (PET) film was increased by about 65% after it was conditioned in CO_2 at ca. 650 psia and 26°C for 24 hours. With such a substantial rise in permeability, it is possible that an additional non-activated transport mechanism became involved as in the case of drawn film described by Matulevicius et al. (1967) and Vieth (1981). That is to say that, due to internal stress induced by conditioning, the average size and/or number of microvoids may increase, facilitating the formation of micro-channels. A completely different set of transport mechanisms would result; e.g., Knudsen flow is a possible additional transport mechanism in this case. Such microstructural gradients are foreshadowed in the quenched melt. Their elimination is facilitated by efficient melt cooling techniques (Owens and Vieth, 1965).

In our laboratory it was observed that the permeation rate of the conditioned film decreased with time after the temperature of the system was raised to 50°C (ca. 30°C below T_g) as also found by Toi in his recent work (1985), while that of the as-received Mylar® film was found to be unchanged. As can be shown in Figure 3.8, the ratio of the permeabilities follows a linear relationship with log of time. Aging under a severe temperature condition (very near T_g) then can eliminate these low activation energy pathways. Thus, one cannot separate aging from previous history since they go hand in hand. Therefore, as far as transport properties are concerned, a mechanistic or model switch during severe sub-T_g annealing is not a farfetched event.

The interplay between conditioning and annealing or aging in varying sample history has important industrial implications. Small gas molecules present in glassy materials at a sufficient level and at an appropriate temperature can act as plasticizers. Under the influence of thermal activation and the "lubricating" effect of volatile plasticizing penetrant molecules, T_g can be substantially reduced as has been

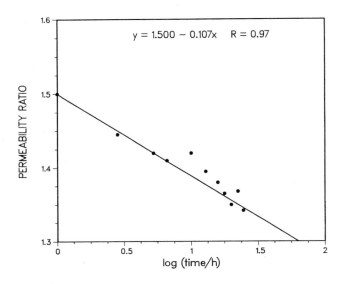

Figure 3.8   Permeability ratio versus time.

observed by Chiou et al. (1985) for a series of polymers; crystallization was induced as the experimental temperature approached and passed T_g. Since PET is rather easily crystallized due to its planar molecular architecture,   substantial increase in crystallinity was observed.   The author and L. Ryder (1983, 1984) also observed an increase in shelf-life for PET containers which had undergone this type of treatment.   Sharp drops in diffusivity and permeability after conditioning near T_g were observed in Toi's work as well.   The possible additional features described above give one pause in approaching the issue of additional transport modeling;   nonetheless it is obvious that the mathematics will become more tractable and capable of refinement   when parametric behavior related to the transport process is better understood.

Since the reduction in the amount of amorphous *trans* is related to the aging process, one might think of the aging, in general terms, as a kinetic process in which isomerization reactions take place in the supercooled liquid.   High energy species is converted to lower energy species in the course of aging.   In the case of PET, the extended *trans* form possesses a higher energy, about 0.52 Kcal/mol more than that of the *gauche* conformer (Koenig and Kormos, 1981).   The rate and the extent of this reaction are related to the excess volume, hence the Langmuir sorption capacity, which is influenced by the excess entropy and history (or initial conditions).   As a matter of fact, just such a molecular approach was taken by Robertson (1979) in modeling the aging process, in which fractional free volume is used to describe the environmental change which imposes a constraint upon the *trans-gauche* isomerization reaction.   This model has succeeded in qualitatively describing the complex volumetric recovery.

Table 3.2   Estimation of Trans Composition in Films

Film Type	Absorbance @973 cm-1	Absorbance @793 cm-1	Absorbance Ratio	*Trans* Percentage
Completely Amorphous	.19397	.32541	0.5961	14.2
Semi-crystalline	.25551	.21924	1.1659	42.5
Unconditioned Mylar®	.30686	.20508	1.4963	59.0
Conditioned Mylar®	.31967	.20747	1.4916	63.0

The modeling aspect of the time-dependent diffusion process in non-equilibrium glassy polymers represents a continuing major challenge.   However, with data in hand on the behavior of the instantaneous diffusivities, one would be in a better position to attack this problem. The mathematical model for diffusion in glassy polymers advanced by Cohen (1983, 1984) caught our attention in that it appears general enough to unify many diverse and complicated behavior patterns of the diffusing penetrant in this class of materials.   Cohen proposed the following model to describe the diffusional transport process in glassy polymers:

$$\frac{\partial C}{\partial t} = \frac{\partial}{\partial x} [D \, \partial C / \partial x] + R [x,t,C,C_t,C_x] \qquad [3.3]$$

$$D = \int_0^t K [t,s,C(x,s), D(x,s)] \, f [C(x,s), C_t(x,s), D(x,s)] \, ds \qquad [3.4]$$

The various possible diffusional behavior patterns can be properly accounted for by choosing the appropriate expressions for R, f and the kernel K, as discussed by Cohen.

3.6 PERMSELECTIVITY OF CELLULOSE ACETATE MEMBRANES

In continuing the cycle of this chapter we turn briefly to a consideration of gas transport in reverse osmosis membranes. Stern and DeMeringo (1978) and Stern and Kulkarni (1982) carried out a series of investigations which employed gas molecules as probes to gain insight into the microstructure of the dense layer of the asymmetric cellulose acetate membrane used in reverse osmosis. Their research also established the permselective character of this membrane in the dry state toward components of gas mixtures.

The solubility of CO_2 was measured at temperatures from 0 to 70°C, while CH_4 measurements were carried out in the range -10° to 30°C. Pressure ranges for both studies ended in the neighborhood of 40-45 atm.

The dual sorption model was obeyed in both cases (up to 60°C for CO_2). The CO_2 work uncovered a relatively large volume of micro-cavities. In the CH_4 studies, a model with two Langmuir sites (Bhatia and Vieth, 1980) best fit the data which also were influenced by a conditioning effect of CO_2 when the polymer was first exposed to high pressure CO_2.

3.7 DUAL SORPTION THEORY APPLIED TO REVERSE OSMOSIS TRANSPORT

In contrast to sorption of simple gases in glassy polymers, sorption of water is often accompanied by significant swelling of the polymer network. Therefore, the concentration of available sites is not fixed, but changes with the degree of sorption. Because the site density is not constant, a single invariant Langmuir model cannot be used to describe this phenomenon. An approach for overcoming this difficulty is taken up next.

The clustering function defined by Zimm and Lundberg (1956) results naturally from the application of statistical mechanics in deciphering the significance of second virial coefficients in the osmotic equation of state for sorbed water:

$$\frac{G_{11}}{\nabla_1} = -\phi_2 \left[\partial (a_1/\phi_1) / \partial a_1 \right]_{P,T} - 1 \qquad [3.5]$$

where G_{11} is the cluster integral, ∇_1 is partial molar volume of water, ϕ_1 and ϕ_2 are volume fractions of water and polymer, respectively, and a_1 is the activity of water.   The quantity $1 + \phi_1 \, [G_{11} \, / \, \nabla_1 \,]$ specifies the average number of water molecules in one cluster;   i.e., it includes the number of molecules in the neighborhood of a reference molecule in excess of random expectation.   Examining eqn. [3.5], it is obvious that this quantity can be obtained from analysis of water vapor sorption isotherm data.

The cluster function can be correlated with pure water permeation rate and sodium chloride rejection (Vieth et al., 1969).   The authors measured the pure water permeation rate, water vapor sorption isotherm and the kinetics of approach to sorption equilibrium for a series of polymers: hydroxyethyl methacrylate (HEMA), copolymers of HEMA and ethyl methacrylate (EMA), cellulose acetate, cellulose nitrate and poly(urethane).   A relationship between the state of water in the membrane and pure water permeability was established.

Following Crank (1956, 1975), typical water vapor sorption kinetics are displayed in Figure 3.9 as fractional equilibration (M_t / M_∞) versus t, where M_∞ and M_t indicate the amounts of water vapor sorbed at equilibrium, and at time t, respectively.

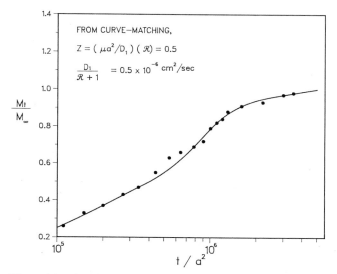

Figure 3.9.   Sorption transient in a poly(urethane).   Poly(ethylene oxide), mol wt 550, 4% cross-linked.

For one-dimensional diffusion, Fick's second law can be modified by including an additional accumulation which accounts for formation of immobilized clusters:

$$D_1 \frac{\partial 2C_1}{\partial x^2} = \frac{\partial C_1}{\partial t} + \frac{\partial S}{\partial t} \qquad [3.6]$$

where S is the concentration of site-bound species (analogous to C_H in the previous analysis). Relaxing the local equilibrium assumption, the rate of formation of immobilized water clusters can be described by a first-order reversible reaction.

$$\frac{\partial S}{\partial t} = \lambda C_1 - \mu S \qquad [3.7]$$

where λ and μ are first-order rate constants for the forward and reverse reactions, respectively. (This approach is somewhat similar to the one of Tshudy and von Frankenberg (1973). However, in their case, the authors considered a reversible site-binding reaction with a fixed number of sites.) At equilibrium,

$$\frac{S}{C_1} = \frac{\lambda}{\mu} \equiv \mathcal{R} \qquad [3.8]$$

When eqns. [3.6] and [3.7] are solved for the case of sorption into a plane sheet from an infinite reservoir, it is found that the effective diffusivity is given by:

$$D_{eff} = \frac{D_1}{\mathcal{R} + 1} \qquad [3.9]$$

As expected, immobilization of the penetrant within the polymer retards the diffusional process by the factor $(1/\mathcal{R} + 1)$. It should be noted that the above development is formally identical to the case of diffusion in a solid coupled with a first order, reversible chemical reaction.

The solution to the above problem has been provided by Crank (1956, 1975) in the form below:

$$\frac{M_t}{M_\infty} = 1 - \sum_{n=1}^{\infty} \frac{2[\mathscr{R}+1+P_n/\mu]^2 \exp[P_n t]}{[\mathscr{R}+1]\{[1+P_n/\mu]^2+R\}k_n^2 a^2} \qquad [3.10]$$

where a represents the membrane half-thickness, $\mathscr{R} = \lambda/\mu$, $k_n = [n+\frac{1}{2}]\,\pi$ and P_n is defined by:

$$k_n^2 = \frac{-P_n}{D_1} \cdot \frac{P_n+\lambda+\mu}{P_n+\mu} \qquad [3.11]$$

By this means, the reduced or *effective* diffusion coefficient of water ($D_1/\mathscr{R}+1$) could be obtained from sorption kinetics. The next order of business was to correlate the above quantity to the quantity $D_1 C_1$ which appears in the equation of Lonsdale, Merten and Riley (1965) for the reverse osmotic permeation of water through the membrane.

$$J_1 = \frac{D_1 C_1 \overline{V}_1}{RT}[(\Delta P - \Delta\pi)/\Delta x] \qquad [3.12]$$

Now, to reiterate briefly, when equilibrium is reached, equation [3.7] becomes:

$$\frac{dS}{dt} = 0 \qquad [3.13]$$

Therefore $\lambda/\mu = S/C_1$. Inserting the latter relation into the quantity $D_1/(\mathscr{R}+1)$:

$$\frac{D_1}{\mathscr{R}+1} = \frac{D_1}{\lambda/\mu+1} = \frac{D_1}{S/C_1+1} = \frac{D_1 C_1}{S+C_1} = \frac{D_1 C_1}{C_0} \qquad [3.14]$$

In other words, the effective diffusion coefficient of water in the membrane is equal to the product of the self-diffusion coefficient of water and the mobile water fraction; i.e., the ratio of the quantity of mobile water to that of total water. In deriving equation [3.12] Henry's law was assumed, but Henry's law is not always applicable. Removing this restriction, and employing D_{eff}, equation [3.12] becomes:

$$J_1 = \frac{D_1 C_1}{C_0} \cdot \frac{\overline{v}_1}{RT} \cdot C_0 [1 + \partial \ln \gamma / \partial \ln C_0]^{-1} [(\Delta P - \Delta \pi)] / \Delta x] \quad [3.15]$$

where γ is the activity coefficient of water in the membrane (i.e., $a = \gamma C_0$). The quantity:

$$[1 + \partial \ln \gamma / \partial \ln C_0]^{-1}$$

covering the swelling effect, can also be obtained from the sorption isotherm. In order to apply it in equation [3.15], the slope of $\ln \gamma$ vs. $\ln C_0$ at $C_0 = 1.0$ has to be obtained.

Thus, the same physical quantity, D_{eff}, is obtained from two independent experiments: pure water permeation in reverse osmosis and water vapor sorption kinetics. From the former experiment the quantity:

$$\frac{J_1 RT}{\overline{v}_1 [(\Delta P - \Delta \pi) / \Delta x] C_0} = \frac{D_1 C_1}{C_0} [1 + \partial \ln \gamma / \partial \ln C_0]^{-1} = D_{eff} \quad [3.16]$$

is obtained and from the latter experiment the quantity:

$$[D_1 / \mathcal{R} + 1][1 + \partial \ln \gamma / \partial \ln C_0]^{-1} = D_{eff} \quad [3.17]$$

is calculated. D_{eff} in both equations has the same physical meaning. Results are shown in Table 3.3 for poly(urethanes) and cellulose nitrate; the estimates of D_{eff} agree reasonably well. Furthermore, the quantity D_{eff} from either equation [3.16] or equation [3.17] was determined with respect to a series of polymers (cellulose acetate, cellulose nitrate, HEMA, HEMA-EMA copolymer and poly(urethane)) and D_{eff} versus cluster size was plotted. The results are shown in Figure 3.10.

Table 3.3   Effective Diffusivities and Cluster Sizes (95% R.H.) at 25°C.

Membrane type	Cluster size at 95% R.H.	Effective diffusivity, cm^2/sec	
		$\dfrac{D_1}{\mathcal{R}+1}[1+\partial \ln \gamma/\partial \ln C_0]^{-1}$	$\dfrac{J_1 RT}{\nabla_1 \dfrac{\Delta P - \Delta \pi}{\Delta x} C_0}$
Poly(urethane), PEO MW 550	6.2	2.3×10^{-6}	6.4×10^{-6}
Poly(urethane), PEO MW 550	5.4	3.9×10^{-6}	3.1×10^{-6}
Cellulose nitrate, unmodified Microporous	0.5	4.8×10^{-7}	2.9×10^{-7}
Cellulose nitrate, 22.3% $TeBr_4$ pre- inclusion	6.3	2.5×10^{-6}	2.3×10^{-6}

The conclusion to be drawn from Figure 3.10 is rather important. Polymeric materials are classified into three regions. Very hydrophilic polymers (on the right side of Figure 3.10) bring about the formation of large water cluster sizes and the diffusion coefficient of water is large. However, the salt rejection is small. This occurs because a large quantity of solvent water is imbibed into the membrane and the polymer is highly swollen. As a result, electrolyte pathways become enlarged and contiguous. Polymers whose hydrophilicity and hydrophobicity are well balanced form smaller cluster sizes. The diffusivity of water in the membrane is also reasonably high (center of Figure 3.10). When the hydrophobicity is high (left side of Figure 3.10) cluster size increases and the diffusivity of water decreases. The large cluster size is caused by trapping of water into pre-existing microvoids in the polymer structure. The path for diffusion from one cluster to another is not contiguous and, as a result, the effective diffusion coefficient of water decreases.

Suppose a large cluster can accommodate electrolyte easily, which in turn results in a low salt rejection; then, change of the nature of the polymer from hydrophilic to hydrophobic first increases the salt rejection, shows a maximum and then salt rejection gradually decreases.

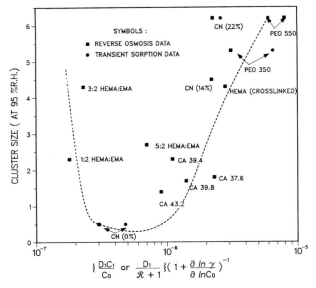

Figure 3.10   Effective diffusivity vs. cluster size.   CA 37.6: Cellulose acetate, 37.6% acetylated; CN(14%): cellulose nitrate, 14% template-expanded;   HEMA: poly(hydroxyethyl methacrylate); 5/2 HEMA/EMA: 5/2 monomer ratio for copolymer of hydroxyethyl methacrylate and ethyl methacrylate;   PEO550: poly(urethane) with poly(ethylene oxide) comonomer of 550 mol wt.

Therefore, the polymeric materials for which hydrophilicity and hydrophobicity are well balanced possess small cluster size and large salt rejection.   Reasonably high water diffusivities in the membrane can also be expected.   This conclusion is the same as the one which was obtained by T. Matsuura (1981) by using a liquid chromatography method.   The cellulose acetate membrane, which opened up the successful practice of the reverse osmosis process, lies precisely in the range where hydrophilicity and hydrophobicity are well balanced.   Both groups' experimental results confirm the above concept.

In the course of our discussion so far in Chapters 2 and 3, three of the four major types of sorption isotherms have been encountered: Henry's law behavior and positive and negative deviations from it.   The isotherm of the biopolymer, collagen, is an example of the fourth type of curve (Figure 3.11).   It passes from a negative to a positive deviation from Henry's law, indicating a transition from site binding at low penetrant levels to diffuse network swelling and cluster formation at high penetrant levels.   The theory presented so far may be applied to the limiting cases of low and high sorption levels, thus bracketing this behavior.

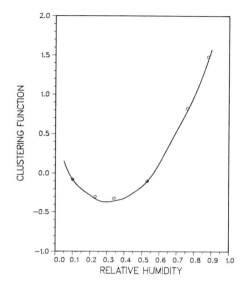

Figure 3.11   Water clustering of glycerine plasticized collagen (30.99% wt/wt) (Lieberman, 1971).

3.8   CONCENTRATION-DEPENDENT TRANSPORT OF GASES AND VAPORS IN GLASSY POLYMERS

At temperatures below the critical for penetrant molecules, the solubility of gases in glassy polymers may sometimes be sufficiently high to require a modification of the dual sorption model.   Stern and Saxena (1980, 1982) modified the partial immobilization model to allow for an exponential concentration dependence of the diffusion coefficient.   This leads to the following expression:

$$\overline{P} = [\, D_o \,/\, \beta\, p\,]\, [\, \exp\,\{\, \beta\, k_D\, p\, [\, (1+FK) \,/\, (1+bp)\,]\, \}^{-1}\,] \qquad [3.18]$$

where D_0 is the diffusion coefficient for the mobile species at infinite dilution and β is a constant which characterizes the concentration dependence of D.

The authors applied this model to a reexamination of the experimental data of Barrer, Barrie and Slater (1958) as shown in Fig. 3.12.   Existence of a minimum in the effective permeability with increasing penetrant pressure was revealed.   In addition, the authors

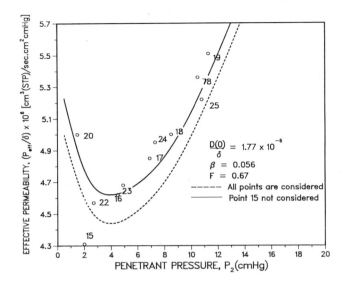

Figure 3.12   Dependence of effective permeability, P_{eff} / δ, on applied penetrant pressure for acetone in ethyl cellulose at 40°C. P_{eff} is the effective permeability coefficient and δ is the mean membrane thickness (Stern and Saxena, 1982).

comment that an indefinite number of essentially reproducible states of such systems, each of considerable lifetime, could be encountered with these history- and time-dependent structures.

3.9 MONOMER LOCALIZATION

Mauze and Stern (1984) analyzed the solubility isotherms for vinyl chloride monomer in glassy poly(vinyl chloride) which were determined by Berens (1978), as exemplified by Fig. 3.13. They concluded that the data could be represented satisfactorily by a dual sorption model composed of a Langmuir expression plus the Flory-Huggins equation (Flory, 1952) to cover ordinary dissolution of the penetrant. The unusual *positive* temperature dependence of one of the model parameters, the hole affinity constant b, suggests the presence of residual solvent or monomer in the polymer samples employed.

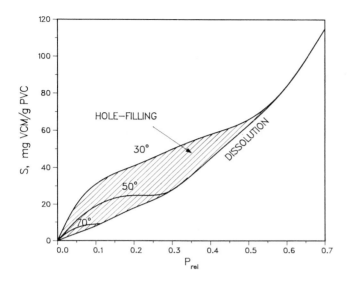

Figure 3.13   Schematic application of dual-mode sorption theory to VCM/PVC solubility isotherms (Berens, 1978).

3.10 CONTROLLED RELEASE
IN VITRO PERMEATION OF SCOPOLAMINE THROUGH HUMAN SKIN

The sorption and rate of permeation of scopolamine base in human skin were measured as a function of drug concentration in aqueous solution contacting the stratum corneum of the skin by Chandrasekaran et al. (1976). The authors observed the nonlinear isotherm and prolongation of the sorption transient characteristic of the dual sorption effect. By applying the total immobilization model, they were able to resolve the isotherm into its components and to explain the diffusional effects. The values of the constants, k_D, C'_H and b were, respectively, 1.1, 5.0 mg/ml and 0.56 mg/ml. (k_D was expressed in the form of a distribution coefficient, C_D / C_{Bulk}, and, hence, is dimensionless in this instance.) They reported "steady state" diffusion coefficients, with a mean value close to 5.0 x 10^{-10} cm^2/sec.

When a second epidermal sample was delipidized, the "steady state" diffusivity rose from 4 x 10^{-10} cm^2/sec to 2 x 10^{-7} cm^2/sec. The authors concluded that the interstitial lipid phase of the stratum corneum acts as the principal permeation barrier. Its removal caused enhancement of the permeation rate by orders of magnitude without causing any change in the equilibrium isotherm, suggesting that scopolamine sorbed by the skin is chiefly localized in the protein phase of the tissue.   The above approach promises to be very useful in predicting drug or toxicant binding and/or diffusibility in human skin.

3.11 MACROMOLECULAR DIFFUSION THROUGH COLLAGEN MEMBRANES

In a recent study (Weadock 1986), collagen membranes prepared by controlled dehydrothermal crosslinking techniques were examined as vehicles for the release of bioactive macromolecules.   The diffusion coefficients for angiotensin II, albumin and aldolase were examined as functions of membrane swelling ratio and diffusant concentration, similar to the work of Hirose et al. (1983) reported in Chapter 9.

Effective diffusivities of the penetrants were found to decrease at low concentrations.   This phenomenon was attributed to adsorption of macromolecules by the collagen membrane.   Bernath et al. (1976) have examined somewhat similar systems in their studies of enzyme therapeutics with collagen-enzyme membrane systems, employing L-asparaginase and urokinase.

Standard free energies of binding, ΔG^0, and the equilibrium binding constant, K_0, were estimated at values of -3 to -9 Kcal/mole and 2×10^2 to 4×10^6 liters/mole, respectively.   Values of K_0 increased several orders of magnitude as the molecular weight of the macromolecule increased from 1040 to 149,100, possibly due to an increase in the number of potential binding sites.   Furthermore, adsorption appeared to be independent of collagen membrane crosslink density, suggesting that sites involved in the crosslinking process are independent of more specific sites involved in *macromolecular* adsorption.   This agrees with the results of Luo et al. (1979), who identified the ϵ-amino group of lysine as the binding locus in collagen for β-galactosidase.

Weadock et al. (1986) have shown that the diffusion of a water-soluble macromolecule (calmodulin, MW=16720) through collagen membranes can be linearly related to the membrane swelling ratio (r*). In their study, high correlation coefficients were obtained for the relationships between r* and D_{eff} for albumin (0.999) and aldolase (0.988).   However, this was not true for angiotensin II, where the correlation coefficient was 0.862.   This indicates that a substantial portion of the diffusion of the relatively low molecular weight angiotensin II occurs through the polymeric collagen network of the membrane as well as through larger aqueous microchannels.

For systems whose behavior can be characterized by dual sorption effects, D_{t1} approaches an asymptotic value D at high source concentration (Michaels et al., 1963). During the process of diffusion, some macromolecules bind to sites available on the collagen membrane. At very low source concentrations, a significant fraction of the diffusing macromolecules are bound to the membrane and as a result, the onset of a steady state flux is retarded.

In these studies, a significant decrease in the value of D_{t1} occurred at low source concentrations (5 ug/ml or less) in almost all time lag experiments. Only in the case of albumin diffusing through crosslinked collagen membranes (for values of r*, the *swelling ratio*, less than 15.35) was this decrease not significant at a 95% confidence interval. These results suggest that diffusion through collagen membranes is consistent with a dual sorption mechanism. One might expect that crosslinking would decrease the number of sites available for sorption to occur, and therefore crosslinked membranes would experience little or no dual sorption effects. However, a decrease in D_{t1} was apparent in both crosslinked and uncrosslinked membranes. This would suggest that the sorption process is independent of sites (charged amino acid residues) involved in the crosslinking reaction. It seems probable that the recent theoretical results of Peppas and Reinhart (1983) ought to be examined for their applicability or possible extension here. This avenue will be pursued in future work.

3.12 ENZYME IMMOBILIZATION BY ADSORPTION

In another recent study, Pedersen et al. (1985) have reported on finite-bath enzyme adsorption in porous supports where a local thermodynamic equilibrium model is employed. They used β-galactosidase and Duolite ion-exchange resin as a model system.

The governing equations in spherical geometry are given as follows:

$$\frac{\partial C_{EL}}{\partial t} + ma_v \frac{\partial C_{ES}}{\partial t} = D \cdot \frac{1}{r^2} \frac{\partial}{\partial r} [r^2 \, \partial C_{EL} / \partial r] \qquad [3.19]$$

where C_{EL} and C_{ES} are the pore liquid and solid (adsorbed) phase enzyme concentrations, a_v is the solid surface area available per unit volume, the void fraction is ϵ_S and the parameter $m = (1 - \epsilon_S) \epsilon_S^{-1}$.

The adsorption equilibrium relationship is of the Langmuir type:

$$C_{ES} = \frac{C'_{ES} \, C_{EL}}{K_S + C_{EL}} \qquad [3.20]$$

While, if the equilibrium assumption is relaxed, the above equation is replaced by:

$$\frac{dC_{ES}}{dt} = k_a \, [C'_{ES} - C_{ES}] \, C_{EL} - k_d \, C_{ES} \qquad [3.21]$$

The reader will readily perceive that these equations are the familiar DST formulation adapted to the case at hand. Convergence of the local equilibrium thermodynamic model with the kinetically-controlled models of other researchers (Bucholz, 1979; Do et al., 1982) has been pointed out for limiting cases by the authors.

Even under long exposure times in the enzyme bath (ca. 10 hrs.), the majority of the enzyme is shown to be confined to the outer half of the support, the loading taking place as a slowly advancing front. The practical importance of considering this type of *anisotropic* distribution of enzyme in reactor design should be clear.

3.13 FACILITATED TRANSPORT

The hierarchy of transport processes encountered in biology begins with passive transport, described by simple Fick's law relationships such as equation [1.1]. At the next level of complexity, facilitated transport appears, in which the steady state penetrant flux is increased through the agency of coupled chemical reactions via an add-on effect. Schultz et al. (1974) describe a number of such cases applying to transport in liquid membranes; in the parlance of biological science the penetrant is "pumped" in the direction of its concentration gradient.

Interestingly, the partial immobilization model for dual mode transport bears a formal analogy to the case of diffusive transport accompanied by equilibrium chemical reaction.

$$J = -D_D \frac{\partial C_D}{\partial x} - D_H \frac{\partial C_H}{\partial x} \qquad [2.22]$$

The flow of the species in the Langmuirian mode $(-D_H \, \partial C_H / \partial x)$ furnishes the add-on effect to the ordinary Fickian transport of the dissolved species v $(-D_D \, \partial C_D / \partial x)$. Because of the reversible site-binding reaction between the penetrant and the membrane, the latter is itself acting as a "pump." Reduced to a permeability relationship (i.e., eqn. [2.24], the term $FK/(1 + bp)$ expresses the facilitation effect.

The change of Langmuirian fluxes in counterdiffusion (section 2.9) has already been described. During the transient, the flux of one species is accelerated by the other and may pass through a maximum. This corresponds to a case of transient facilitation.

In the case of enzymatic reaction, facilitation is accomplished through sharpening of the gradient, $-\partial C_D / \partial x$. The time to achieve steady state (related to Θ, the time lag) is correspondingly reduced, as will be discussed in detail in Chapter 4. In Chapter 9, facilitation of the transport of acetylcholine via the action of acetylcholinesterase in the synaptic cleft-post synaptic membrane microstructure is discussed. The coupled action of receptor-operated ion channels is likewise treated in detail.

The top rung of the transport ladder is occupied by active transport; here a penetrant may actually flow uphill in the sense of its concentration gradient across a membrane. But the process just described is responding to an electrochemical potential gradient as influenced by a complementary phenomenon, such as a membrane pH gradient. This type of transport is highlighted in Chapters 7 and 8 which focus on transport of the inducer, lactose, in the biosynthesis of the enzyme, β-galactosidase, via the lac operon.

3.14 MODELING GAS TRANSPORT IN PACKAGING APPLICATIONS

Recently, the use of thermoplastics as carbonated beverage and aerosol containers has become widespread. These thermoplastics are, in general, either glassy amorphous or semi-crystalline materials. The transport of gases in these polymers, therefore, would be expected to exhibit the dual sorption behavior. Fenelon (1973) has utilized DST to predict the time dependence of pressure loss in pressurized plastic containers; he also carried out a series of experiments to estimate the parameters for a pair of candidate materials.

Fenelon showed that three material parameters, namely: gas sorption, gas permeation and creep, influence the rate and magnitude of pressure loss in a pressurized plastic container. Theoretical expressions have been developed allowing individual calculation of each of these contributing factors from readily obtained laboratory test data. A hypothetical carbonated plastic beverage container was assumed to be equivalent to an open-ended cylinder. The experimental parameters which define carbon dioxide transport in poly(ethylene terephthalate) film at 40°C, as described in the work of Vieth and Sladek (1965), were adopted by Fenelon. He assumed that the influence of the container liquid environment on material transport properties could be safely neglected. The pressure is considered uniform throughout the internal volume of the container and thereafter continuously decreases from an initial value, p_0, as it enters the container walls whose initial concentration of carbon dioxide is zero. The percentage pressure drop as a function of time as defined in the following equation was calculated:

$$\% \text{ pressure drop } (PD)_t = [p_0 - p_t / p_0 - p_f] \times 100\% \qquad [3.22]$$

where p_0 is the initial pressure, p_t is the pressure at time t, and p_f is the final pressure, in this case, 1 atm. Figure 3.14 shows the results of Fenelon's calculation. Curve A represents total pressure drop, i.e., total gas loss by permeation and sorption (or immobilization). Curve B represents pressure loss by sorption only. Curve C represents pressure loss by permeation, i.e., normal diffusion. It is interesting to note that pressure loss at short times is primarily dependent on gas sorption. As time proceeds, the permeation contribution, curve C, becomes important. This is consistent with physical reasoning. One could imagine that, at the beginning stage of transport, the microvoids or holes are filling with the gas molecules. After completion of this immobilizing process, the normal diffusion process governs.

Pressure loss calculated from the numerical solution to the total immobilization model was compared with the pressure drop calculated from the steady state permeability analysis. Figure 3.15 shows such a comparison. It is illustrated that for a maximum of 10% pressure drop,

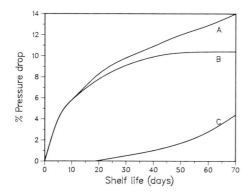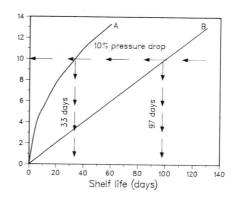

Figure 3.14 Pressure drop vs. shelf life (Fenelon, 1973). Curve A: total pressure drop; curve B: pressure drop by sorption; curve C: pressure drop by permeation.

Figure 3.15 Percent pressure drop vs. shelf life for a pressurized hypothetical container based on a nonsteady state (curve A) and steady state (curve B) analysis, respectively (Fenelon, 1973).

the classical steady state analysis predicts a shelf life which is about three times longer than that predicted from an unsteady state analysis. This means that the actual gas depressurization is three times larger than the predicted pressure loss estimated when the steady state analysis is used alone. This is a significant demonstration of the importance of considering dual sorption. In designing a carbonated beverage container using glassy polymeric materials, an unsteady state analysis in predicting the performance of the material is mandatory.

In view of the above findings, Fenelon (1973) further examined the transport of carbon dioxide in candidate glassy polymers. He experimentally verified the utility of the dual sorption theory by studying its capability of predicting steady state dynamic transport parameters from equilibrium and transient sorption behavior. Two polymeric materials, Mylar® Type A (PET) and an experimental High Acrylonitrile (HAN) polymer, were chosen. These materials are of significant industrial importance in packaging applications. Initial, transient, and equilibrium sorption levels were measured on a sensitive pressure transducer with a range of 0-300 psi absolute pressure. Steady state CO_2 and O_2 permeation coefficients were measured on a modified Dow cell at 25°C. Pressure differentials, which covered a range from 1 to 5 atm were used. The equilibrium isotherms for these CO_2-polymer systems were found to be for Mylar® Type A,

$$C = \frac{6.0 \times 0.37\ p}{1 + 0.37\ p} + 0.14\ p \qquad [3.23]$$

and for the experimental HAN polymer,

$$C = \frac{6.0 \times 0.55\ p}{1 + 0.55\ p} + 0.49\ p \qquad [3.24]$$

The same technique as described previously was used to calculate the mobile species diffusivity. The calculated values of diffusivities were 3.76×10^{-9} and 2.0×10^{-10} cm^2/sec for Mylar® Type A and experimental HAN polymers, respectively. The permeability coefficients were then calculated using $\overline{P} = k_D \cdot D$, where k_D is the Henry's law constant from the dual sorption analysis.

For CO_2 permeation at 25°C, Mylar® Type A polymer,

Permeability $(\overline{P})$ = $0.14 \times 3.76 \times 10^{-9} \times (2.19 \times 10^{10})$
= 11.5 cc mil/24 hr atm 100 in^2

For CO_2 permeation at 25°C, experimental HAN polymer,

Permeability $(\overline{P})$ = $0.49 \times 2.0 \times 10^{-10} \times (2.19 \times 10^{10})$
= 2.2 cc mil/24 hr atm 100 in^2 ,

where 2.19×10^{10} is a conversion factor.

The permeability coefficients $(\overline{P})$ obtained from steady state permeation rates measured with a Dow cell were found to be 12.7 (as compared with 11.5) for Mylar® Type A polymer and 2.6 (as compared with 2.2) for experimental HAN polymer. The observed agreement between measured (steady permeation) and calculated (dual sorption transient analysis) values is rather good, albeit consistently a little on the low side for the calculated values. This may possibly be due to neglect of Langmuir mode transport. As pointed out by Fenelon, these values are within experimental confidence limits of the various test procedures used. Masi and Paul (1982) have subsequently carried out simulations using the fully developed dual sorption, dual mobility model. A comparison of their results with those given by Fenelon (1973) is shown in Fig. 3.16. The slightly lower total CO_2 loss predicted by assuming total immobilization is reasonable since, as

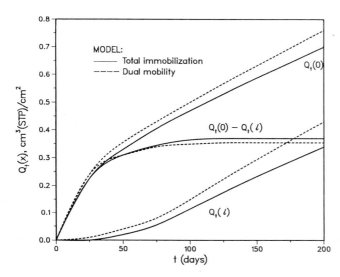

Figure 3.16   Comparison of total immobilization model prediction with that obtained using dual mobility model (Masi and Paul, 1982).

mentioned, this model neglects the additional parallel mode of transport provided by the Langmuir term.

3.15 PERSPECTIVE ON MEMBRANE SEPARATIONS PROCESS PRINCIPLES AND SCALE-UP PARAMETERS

Membrane processes such as gas permeation, liquid permeation, and reverse osmosis depend on a solution transport mechanism in which transport rates can be expressed in terms of a diffusion constant and the activity gradient.   Transport rates are related in an important way to interactions between the penetrants and membrane, the size and shape of the penetrating molecules, and membrane microstructural features such as free volume occupancy and/or swelling effects.   As in reverse osmosis, it is possible to take advantage of the increase in solvent activity with increased pressure to enhance driving forces.   In contrast, in ultrafiltration, the importance of transport by activated diffusion within the solid is greatly diminished by parallel phenomena such as flow in pores.

Translating the fundamental principles outlined so far into an industrial   separation   process   requires   a   variety   of   chemical engineering concepts.   In many ways, such concepts are analogous to those used in a number of more familiar operations.   Similar to relative volatility in distillation, the separation factor for one component relative to another is expressed by the permselectivity, defined as the ratio of permeation fluxes of the two components as measured by permeation of the mixture.

For dilute solutions of fluid in the membrane, exemplified by fixed gases, the permselectivity is equal to the ratio of the permeabilities of the pure components.

$$\alpha_{AB} = \frac{\overline{P}_A}{\overline{P}_B} \qquad\qquad [3.25]$$

Thus, the permselectivity is predictable over the entire composition range in these cases; e.g., the permeation of binary gas mixtures in separation units employing glassy polymer membranes.   Because intact glassy polymer membranes are highly discriminatory toward small molecule penetrants, ratios of permselectivities toward mixtures of, say, H_2/CO_2 can be quite favorable (Sengupta and Sirkar, 1984).   For fluids with high solubilities in the membrane, in excess of perhaps several percent, interactions between fluid components and polymer-solvent interactions become important, affecting the separation factor so that the simple ratio of pure-component fluxes is no longer applicable.

In addition to having available a wide variety of film-forming synthetic plastics to provide a broad selection of membrane materials, processes may also employ membranes which have been "tailored" to improve their properties vis-a-vis the desired separation.   In a sense, the developments of Loeb and Sourirajian with cellulose acetate RO membranes (1962) represent the results of membrane tailoring studies, where the polymer is cast in such a form (asymmetric membrane with ultrathin barrier layer) as to optimize its use as a separation barrier. In addition to improving macrophysical properties, such as effective thickness, techniques have been employed to alter both the chemical structure of the membrane material and its microphysical properties.

Where complete freedom of choice is available, economic factors will determine the balance to be struck between high permeability and high selectivity for any given separation.   Many separations regarded as candidates for membrane systems would likely be of mixture components that are rather closely related chemically, and which would therefore have relatively modest permselectivities.   As a result, the membrane unit would of necessity be a staged system, and thus the permselectivity becomes important in determining the number of stages in the unit. Power must be expended for recompression of permeated fluids, and this will be multiplied severalfold in a multistage unit. Balanced against this, high permeability will mean decreased membrane surface area requirements, and a decreased investment not only in membranes but in equipment housing them. Thus, a balance must be achieved between power costs (and the investment in pumps and compressors) vs. capital expenditure for membranes and associated equipment.

In reviewing the engineering considerations that must be taken into account in the design of a membrane separation unit, it is of interest first to compare aspects of the process with established separations involving mass transfer between phases. Consider, for example, the comparison between solvent extraction, where liquid-phase diffusion controls, and separation of a liquid mixture with a semipermeable membrane. In both cases, the steady state mass transfer capacity per unit equipment volume can be expressed to a first approximation by an equation of the following "film theory" form:

$$N_A = D_A a(C_{A_2} - C_{A_1})_{mean/\ell} = k_L a (C_{A_2} - C_{A_1})_{mean} \qquad [3.26]$$

N_A is the mass flux per unit volume, D_A the suitable concentration-mean, or effective, diffusivity, a the transfer surface area per unit volume, C_{A_1} and C_{A_2} are concentrations, and ℓ the "effective" film or membrane thickness. (In the case of solvent extraction, D_A, ℓ and a are usually lumped into the factor $k_L a$.) Comparing each of these quantities for the two types of processes, the major difference appears to be in the diffusivity. Liquid phase diffusivities are perhaps two orders of magnitude higher than those for a liquid permeant in a polymeric membrane. Cognizance of this obvious difference has probably been the principal obstacle hindering more rapid development of membrane pervaporation processes. However, it is pertinent to consider the other factors in the transfer rate equations. The mean driving forces, expressed as $\triangle C_A$, are of the same order of magnitude. In the case of extraction, this represents the difference between the local bulk stream concentration and the concentration in equilibrium with the absorbing solution. On the other hand, liquid "film thicknesses" in extraction are of the order of one millimeter, or 40 mils, while membranes of good mechanical integrity and stability can be prepared at thicknesses of one mil, a difference of about an order of magnitude.

Thus, since the capacities per unit volume achievable in extraction equipment are representative of commercial possibility, an additional order of magnitude at least must be made up in membrane systems by increased transfer surface per unit volume. Since typical packed beds give values of "a" of about 10 to 50 ft^2/ft^3 this means that membrane geometries must be conceived to give surface-to-volume ratios of at least several hundred or perhaps 1,000 ft^2/ft^3.

Indeed, publications and patents describe designs that approach or exceed this figure (see Table 3.4). Such designs involve either stacks of thin membranes uniquely spaced and manifolded, or spirally wound membranes, as for instance reported by Michaels, Vieth and Bixler (1963), fitted with spacers to allow for fluid contact on either side. If the edges are suitably glued and a perforated central collecting tube is added, one has the typical spiral wound reverse osmosis module (see Fig. 3.17). Under these circumstances, then, it appears that satisfaction of an important criterion for commercial feasibility is at hand.

At least two problems are associated with the use of a complex geometry to generate high membrane surface areas. First, problems of fluid baffling and manifolding can lead to high fabrication costs, which may add further expense to an already costly membrane requirement. Secondly, particularly where gases or vapors are being separated, power requirements merely to overcome pressure drop across the upstream face of the membranes may become a significant operating expense, due to restrictions in flow paths caused by intricate baffling and membrane design.

An additional problem that can occur with improper design of the system is concentration polarization at the membrane surface. Most pronounced in some liquid phase systems, where transport rates through the membrane begin to approach those of the components in the liquid phase, an additional mass transfer resistance can build up at the membrane surface, decreasing both the total permeation throughput and the achievable separation. Experimental studies of dialysis membranes in the artificial kidney have shown that liquid-phase mass transfer can offer a substantial resistance unless special measures are taken, such as the use of very narrow flow channels. Similarly, studies of sea-water desalination have shown that some degree of concentration polarization is unavoidable with the development of high-flux membranes.

While the comparison given above between solvent extraction and membrane separation may place due emphasis on the important factors in gaining high capacity per unit volume, close correspondence breaks down when comparing the separation attainable with each system. Solvent extraction has the advantage of multistage operation through countercurrent action, while a membrane process must be laid out in

Table 3.4. Commercial Membrane Modules, their Properties and Application in Biotechnology (Strathmann, 1985)

Module type	Membrane area per module volume (m^{-1})	Investment costs	Operating costs	Control of concentration polarization and membrane fouling	Area of application
Tubular	20-30	very high	high	very good	Cross-flow filtration, ultrafiltration of solutions with high solids content
Plate-and-frame	400-600	high	low	fair	Cross-flow filtration, ultrafiltration, pervaporation
Spiral wound	800-1000	low	low	poor	Reverse osmosis, ultrafiltration of solutions showing low tendency of membrane fouling
Capillary	600-1200	low	low	good	Ultrafiltration and pervaporation
Hollow fiber	- - -	very low	low	very poor	Reverse osmosis of relatively 'clean' solutions, gas separation

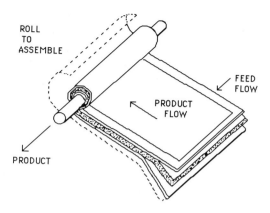

ROLL
TO
ASSEMBLE

FEED
FLOW

PRODUCT
FLOW

PRODUCT

Figure 3.17 Spiral wound module.

discrete stages.   Qualitatively, balancing this, however, is the fact that the solid membrane often acts as a superior separation barrier compared to the liquid-liquid phase boundary.

Design of a staged membrane separation unit can successfully employ cascade concepts, developed originally for isotope separations. Fortunately, because of the sizable effect on permeation rates of small differences in molecular structure, liquid permselectivities are far more favorable than typical isotope separation factors.   Thus, it appears possible that many difficult separations could be accomplished over only a few stages in a membrane permeation unit.

The "ideal cascade" is based on the equality of composition of recycle and fresh feed streams to each stage in the cascade.   This condition leads to the minimum separation duty for the unit, and results in fewer stages than the amount necessary to accomplish a given separation with the standard constant reflux-ratio design employed in fractionation towers.   In the latter, constant reflux allows the heat removal to occur at a single location.   In the tapered ideal cascade (variable reflux ratio), the energy input is regulated at each stage, thus providing the opportunity for optimum control of the design.

A second unique aspect of a cascaded membrane system is that theoretical stage efficiency approaches 100%.   That is to say, barring concentration polarization, there is little opportunity for loss of separation efficiency across the membrane barrier.   The analog of entrainment would be leakage of feed into product due to membrane defects, which can be obviated. An example of a clever solution is one described by Henis and Tripodi (1981), involving a composite membrane having a thin film of silicone rubber covering the defects in a glassy polysulfone membrane. Thus, in comparing a membrane separation with other types of staged processes through consideration of separation factors, the membrane system has a possible advantage in lower actual stage requirements over those separations where stage efficiencies are some fraction of theoretical. Secondly, this stage efficiency allows design with a high degree of accuracy, since conventional empirical efficiency factors may not be as important.

It is clear then that the application of known chemical engineering principles to the design of a membrane separation unit can result in considerable economies relative to fabricating membranes and membrane stages in configurations which are novel or are borrowed from other separation processes. The combination of suitable design concepts with the variety of membrane materials available today, plus known techniques for altering membrane structure to improve selective properties, has proven capable of bringing the process into competition with established separation methods. One may confidently anticipate that many successful applications are soon to come onstream. For example, Monsanto markets Prism® membrane gas separators for hydrogen recovery from ammonia purge streams, carbon dioxide recovery for reinjection in enhanced oil recovery operations and dehydration of air (C&EN, 1985). In addition, Monsanto and Dow have developed units to produce on-site, low cost nitrogen gas based on removal of oxygen and water vapor from air through hollow fiber membranes. The Monsanto unit is based on the firm's polysulfone hollow fiber membranes used in gas separations. Dow's units use a proprietary polyolefin membrane (C&EN, 1985).

3.16 REFERENCES

Anon., *C&EN*, **Mar. 4**, 26 (1985).
Anon., *C&EN*, **Feb. 18**, 35 (1985).
Barrer, R.M., J.A. Barrie and J. Slater, *J. Polym. Sci.*, **27**, 177 (1958).
Berens, A.R., *J. Mem. Sci.*, **3**, 247 (1978).
Bernath, F.R., L.S. Olanoff and W.R. Vieth, in "Biomedical Applications of Immobilized Enzymes and Proteins," T.M.S. Chang, ed., Plenum Press: New York, 351 (1976).

Bhatia, D. and W.R. Vieth, *J. Mem. Sci.*, **6**, 351 (1980).

Bucholz, K., *Biotechnol. Lett.*, **1**, 18 (1979).

Chan, A.H. and D.R. Paul, *J. Appl. Polym. Sci.*, **24**, 1539 (1979).

Chan, A.H. and D.R. Paul, *J. Appl. Polym. Sci.*, **25**, 971 (1980).

Chan, A.H. and D.R. Paul, *Polym. Eng. Sci.*, **20**, 87 (1980).

Chandrasekaran, S.K., A.S. Michaels, P.S. Campbell and J.E. Shaw, *AIChE J.*, **22**, 828 (1976).

Chen, S.P., *Am. Chem. Soc., Div. Polym. Chem. Polym. Prepr.*, **15**, 77 (1974).

Chen, S.P., *Polym. Eng. Sci.*, **21**, 922 (1981).

Chiou, J.S., J.W. Barlow and D.R. Paul, *J. Appl. Polym. Sci.*, **30**, 3911 (1985).

Cohen, D.S., *J. Polym. Sci.*, **21**, 2157 (1983).

Cohen, D.S., *J. Polym. Sci.*, **22**, 1001 (1984).

Crank, J., "The Mathematics of Diffusion," Oxford University Press: London 1956).

Crank, J., "The Mathematics of Diffusion," 2nd ed., Clarendon Press: Oxford (1975).

Dao, L., Ph.D. Thesis in Chemical and Biochemical Engineering, Rutgers University (1987).

Do, D.D., D.S. Clark and J.E. Bailey, *Biotechnol. Bioeng.*, **24**, 1527 (1982).

Eilenberg, J.A. and W.R. Vieth, *J. Appl. Polym. Sci.*, **16**, 945 (1972).

Fenelon, P.J., *Polym. Eng. Sci.*, **13**, 440 (1973).

Flory, P.J., *J. Chem. Phys.*, **20**, 51 (1952).

Heffelfinger, C.J. and P.G. Schmidt, *J. Appl. Polym. Sci.*, **9**, 2661 (1965).

Henis, J.M.S. and M.K. Tripodi, *J. Mem. Sci.*, **8**, 233 (1981).

Hirose, S., W.R. Vieth and M. Takao, *J. Molec. Catal.*, **18**, 11 (1983).

Ito, E., K. Yamamoto and Y. Kobayashi, *Polymer*, **19**, 39 (1978).

Koenig, J.L. and D.E. Kormos, *N.Y. Acad. Sci.*, **371**, 87 (1981).

Koros, W.J. and D.R. Paul, *J. Polym. Sci., Polym. Phys. Ed.*, **16**, 1947 (1978).

Koros, W.J. and D.R. Paul, *Polym. Eng. Sci.*, **20**, 14 (1980).

Kovacs, A.J., *J. Polym. Sci.*, **30**, 131 (1958).

Levita, G. and Th.L. Smith, *Polm. Eng. Sci.*, **21**, 936 (1981).

Lieberman, E.R., Ph.D. Thesis in Food Science, Rutgers University (1971).

Lin, S.B. and J.L. Koenig, *J. Polym. Sci., Polym. Phys. Ed.*, **20**, 2277 (1982).

Loeb, S. and S. Sourirajan, *Adv. Chem. Ser.*, **38**, 117 (1962).

Lonsdale, R.K., U. Merten and R.L. Riley, *J. Appl. Polymer Sci.*, **9**, 1341 (1965).

Luo, K.M., J.R. Giacin, S.G. Gilbert and E.R. Lieberman, *J. Molec. Catal.*, **5**, No.1, 15 (1979).

Masi, P. and D.R. Paul, *J. Mem. Sci.*, **12**, 137 (1982).

Matsuura, T., "Fundamentals of Synthetic Membranes," Kitani Shobo Co.: Tokyo p.151 (1981) .

Matulevicius, E.S., W.R. Vieth and S.R. Mitchell, *Kolloid Zeitschrift und Zeitschrift fur Polymere*, Band 220, Heft 1, Seite 49 (1967).

Mauze, G.R. and S.A. Stern, *J. Mem. Sci.*, **18**, 99 (1984).

Meares, P., *J. Am. Chem. Soc.*, **76**, 3415 (1954).

Michaels, A.S., W.R. Vieth and J.A. Barrie, *J. Appl. Phys.*, **34**, 1 (1963).

Michaels, A.S., W.R. Vieth and H.J. Bixler, *Polymer Letters*, **1**, 19 (1963).

Moore, R.S., J.K. O'Loane and J.C. Shearer, *Polym. Eng. Sci.*, **21**, 903 (1981).

Owens, J.E. and W.R. Vieth, U.S. Patent Number 3,223,757, Process for Quenching an Extruded Polymer (1965).

Pedersen, H., L. Furler, K. Venkatasubramanian, J. Prenosil and E. Stuker, *Biotechnol. Bioeng.*, **27**, 961 (1985).

Peppas, N.A. and C.T. Reinhart, *J. Mem. Sci.*, **15**, 275 (1983).

Petrie, S.E.B., "Polymeric Materials: Relationships Between Structure and Mechanical Behaviors," Am. Soc. for Metals: Metal Park, Ohio (1975).

Robertson, R.E., *J. Polym. Sci., Polym. Phys. Ed.*, **17**, 597 (1979).

Sengupta, A. and K.K. Sirkar, *J. Mem. Sci.*, **21**, 73 (1984).

Schultz, J.S., J.D. Goddard and S.R. Suchdeo, *AIChE J.*, **20**, 417 (1974).

Smith, Th.L, *Polym. Eng. Sci.*, **21**, 903 (1983).

Stern, S.A. and A. H. DeMeringo, *J. Polymer Sci.*, **16**, 735 (1978).

Stern, S.A. and S.S. Kulkarni, *J. Mem. Sci.*, **10,** 235 (1982).

Stern, S.A. and V. Saxena, *J. Mem. Sci.*, **7**, 47 (1980).

Stern, S.A. and V. Saxena, *J. Mem. Sci.*, **12**, 65 (1982).

Strathmann, H., *Trends in Biotechnol.*, **3 No. 5**, 112 (1985).

Struik, L.C.E., "Physical Aging in Amorphous Polymers and Other Materials," Elsevier: Amsterdam (1978).

Tant, M.R. and G.L. Wilkes, *Polym. Eng. Sci.*, **21**, 874 (1981).

Toi, K., T. Ito and I. Ikemoto, *J. Polym. Sci., Polym. Let. Ed.*, **23**, 525 (1985).

Tshudy, J.A. and C. von Frankenberg, *J. Polym. Sci.*, Part A-2 **11**, 2027 (1973).

Vieth, W.R., A.S. Douglas and R. Bloch, *J. Macromol. Phys.*, **B3**, 737 (1969).

Vieth, W.R., J.M. Howell and J.H. Hsieh, *J. Mem. Sci.*, **1**, 177 (1976).

Vieth, W.R. and K.J. Sladek, *J. Coll. Sci.*, **20**, 1014 (1965).

Vieth, W.R., *Fifth Annual Ryder Conf. on Oriented Plastic Containers* (1981).

Vieth, W.R. and L. Ryder, U.S. Patent Number 4,426,427, Presaturated Preform (1983).

Vieth, W.R. and L. Ryder, U.S. Patent Number 4,435,453, Presaturated Hollow-plastic Article (1984).

Weadock, K., Ph.D. Thesis in Biomedical Engineering, Rutgers U. (1986).

Weadock, K., F.H. Silver and D. Wolff, *Biomaterials*, in press (1987).

Wonder, A.G. and D.R. Paul, *J. Membr. Sci.*, **5**, 63 (1979).

Zimm, B.H. and J.L. Lundberg, *J. Phys. Chem.*, **60**, 425 (1956).

Sometimes I'm home in New
Jersey,
Sometimes I'm overseas.
Sometimes I like to go fishin',
Sometimes with the Japanese.

4

MEMBRANE BIOSENSORS

4.0 INTRODUCTION

We must begin now in earnest to cross the bridge to the province of the biological systems. Our laboratory entered this field in the latter part of the 1960s; in the early 1970s, we had the good fortune to begin a collaboration with the laboratory of Drs. Shuichi Suzuki and Isao Karube at the Tokyo Institute of Technology; a laboratory which has gained distinction for its work on membrane biosensors.

Dr. Sachio Hirose is a graduate of the program at Tokyo Institute of Technology who worked in our laboratory on membrane laminate models of the neuromuscular junction, as described in Chapter 9 of this book. He is with Mitsubishi Petrochemical Company, Ltd., his employer since graduation. Recently he made a compilation of the scientific information pertaining to biosensor membranes used for analysis. As part of our continuing collaboration, the descriptions which form the first part of this chapter are based largely on Dr. Hirose's work. The mathematical analyses which follow are derived from work in our laboratory.

4.1 BIOSENSOR MEMBRANES FOR ANALYSIS

In recent years, the development of biofunctional sensors capable of detecting and analyzing a variety of chemical components has been achieved (Suzuki, 1981, 1983, 1984; Rechnitz, 1984; Seitz, 1984; Wohltjen, 1984; Arnold and Meyerhoff, 1984). Polymer membranes are utilized as carriers for active biological entities which serve as the detection elements in these sensors. Enzyme membranes formed the first generation of such devices. Currently, novel membranes are being designed and perfected in which living organisms have been immobilized in/on polymer membranes. Taken as a whole, these developments comprise the family referred to as *biosensor membranes*.

4.2 TYPES OF BIOSENSOR MEMBRANES (BM)

As shown in Figure 4.1, the biosensor membrane is typically a product formed via a process in which a molecular identification element, such as an enzyme, has been immobilized in/on a natural or synthetic polymer membrane. Chemical substances (substrates) enzymatically react within the biosensor membranes. Electrode-sensitive substances produced or consumed as a result of the reaction diffuse into the BM and reach the electrochemical device which transduces the chemical signal. In the electrochemical device, the output electrical signal corresponding to the concentration of a substance is obtained. In this manner, the concentration of the chemical substance can be rapidly and conveniently detected.

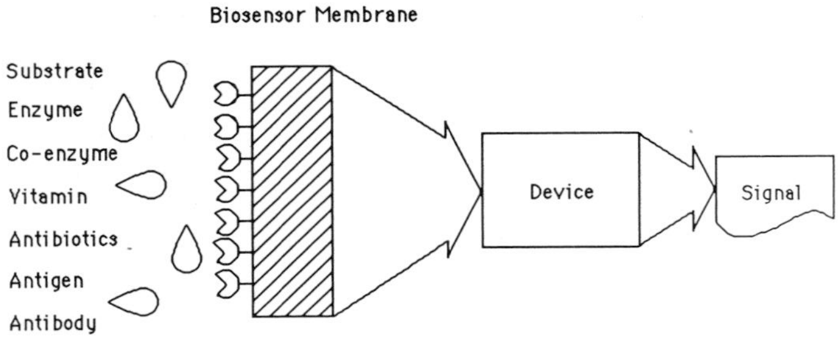

Figure 4.1   Varieties of biosensor membrane applications.

As shown in Figure 4.2, a variety of biosensor membranes have been designed using enzymes, bacteria, organella, antibodies, antigens, receptors, tissues of animals and plants, etc. Recently, these molecular identification elements have been combined (Pau and Rechnitz, 1984) and hybridized (Blum and Coulet, 1984; Schubert et al., 1984; Karube, 1983), as well as employed in an amplifying biosensor (Schubert et al., 1985; Ikariyama et al., 1985). In addition, there have appeared the semiconductor biosensor in which a semiconductor replaces the usual electrode; the thermistor-enzyme biosensor; the piezobiosensor

(Guilbault, 1983); and an optobiosensor using light fibers (Arnold, 1985).

ENZYME-BIOSENSOR MEMBRANES (EBM)

The enzyme-biosensor (enzyme electrode) is a prototype of the biosensor proposed by Updike and Hicks (1967). Figure 4.3 illustrates one example of an enzyme-biosensor, the glucose sensor. Ultrafiltration membranes (to exclude protein) remove the macromolecular substances present in the solution and permit the permeation of the glucose substrate (glu), and oxygen. In the EBM, the enzyme reaction takes place as shown below:

$$\text{glucose oxidase}$$
$$\beta\text{-d glucose} + O_2 \quad \rightarrow \quad \text{gluconolactone} + H_2O_2$$

Unconsumed oxygen diffuses through the gas-permeable membranes and is converted into an electrical signal by the oxygen electrode. The substances (e.g., O_2, H_2O_2, CO_2, NH_3, etc.) consumed or produced by the EBM are measured by the variations in current values, effected by the oxygen electrode, or by variations in potential difference, effected by the ammonia gas electrode.

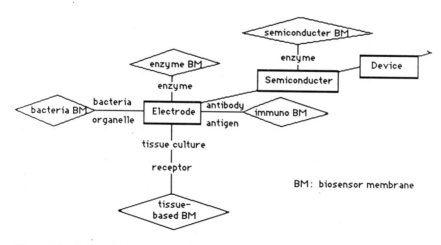

Figure 4.2   Types of biosensor membranes.

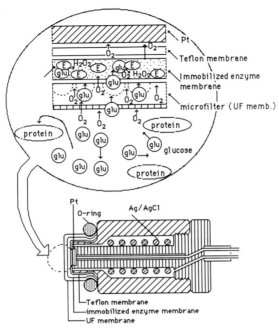

Figure 4.3   Glucose biosensor.

BACTERIAL BIOSENSOR MEMBRANES (BBM)

Bacterial biosensors are those sensors which directly utilize a train of bacterial-enzyme reactions to achieve detection (Nikolelis, 1985). The principle is the same as that of the enzyme biosensor, and the substances produced or consumed by the BBM are measured electrolytically so that the concentration of substrate in the specimen is measured. Figure 4.4 illustrates an example of the BBM preparation method, using agar-gel. These membranes have been manufactured and used industrially in the fermentation field to measure amino acids present in a fermentation tank, as well as organic acids, alcohol, etc. Also, the biological oxygen demand (BOD) sensor has entered industrial practice recently.

IMMUNO-BIOSENSOR MEMBRANES (IBM)

In this technique (Haga et al., 1980; Boitieux and Thomas, 1984; Keating and Rechnitz, 1984), an IBM made of an immobilized antibody, a serum-containing antigen and a labeled antigen are forced to come into contact, so that the antigen present in the serum and the labeled antigen will combine competitively with the IBM.   As a result, the membrane

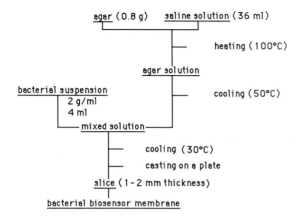

Figure 4.4   Preparation of an agar-bacterial biosensor membrane.

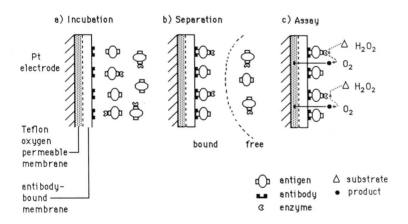

Figure 4.5   Example of an immuno biosensor membrane.

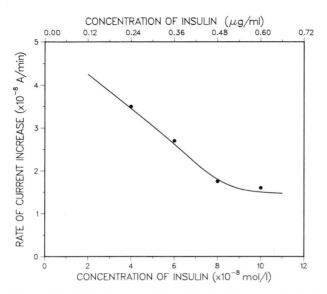

Figure 4.6 Calibration curve for an insulin biosensor. Incubation: pH 8.0, 30°C, 10 min. Assay: in 10 mmol/l H_2O_2 at pH 7.0.

electrical potential, which depends on the quantity of antigen present in the serum, can be detected.   Figure 4.5 illustrates the membrane formula of IBM, and Figure 4.6 illustrates the example of measuring insulin with an IBM containing immobilized insulin antiserum.   Very recently, Bush and Rechnitz (1987) have described the case of membrane electrodes which incorporate antigenic ionophores.

At present, the enzyme immunoassay (EIA) is beginning to compete with the radio immunoassay (RIA), in clinical analysis. However, the present sensitivity of the EIA is far inferior to that of the RIA.   Further development of the EIA as a new measuring system to take the place of RIA may be anticipated in the future through improvements in the selectivity of antibodies, leading to higher sensitivity of the IBM.

TISSUE-BASED BIOSENSOR MEMBRANES (TBM) AND RECEPTOR BIOSENSOR MEMBRANES (RBM)

Rechnitz et al. have designed a TBM comprised of immobilized mammalian tissue and have demonstrated that adenosine, adenosine-5'-1 phosphoric acid, guanine and glutamine can be measured by it (Arnold and Rechnitz, 1981a, b; 1982; 1980).   The authors have also

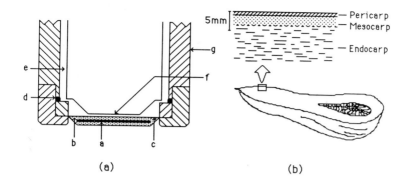

Figure 4.7a   Schematic diagram of the squash tissue-based biosensor membrane.   a) slice of yellow squash tissue;   b) BSA conjugate layer;   c) carbon dioxide gas-permeable membrane;   d) O-ring;   e) internal electrolyte solution;   f) pH-sensing glass membrane;   g) plastic electrode body.

Figure 4.7b   Cross-section of yellow squash showing origin of mesocarp biocatalytic layer.

proposed an IBM made of an immobilized plant tissue slice (e.g., yellow squash (Kuriyama and Rechnitz, 1981), corn (Kuriyama et al., 1983)). Fig. 4.7 illustrates one example of a plant-TBM, and a candidate reaction is shown by the formula below.

$$\text{L-glutamic acid} \quad \xrightarrow[- CO_2]{\text{glutamate decarboxylase}} \quad \gamma\text{-aminobutyric acid}$$

Another development is the formulation of an RBM through the isolation and reconstitution of acetylcholine receptor from the electroorgan of the torpedo fish (Hirose and Vieth, 1984).   Figure 4.8 shows the comparison of the sectional structure of the RBM with a synaptic membrane.   (This is discussed in more detail in Chapter 9.)

Schubert et al. (1982) have formulated a BM made of gelatin-entrapped rat liver cells and have demonstrated that NADH is measurable with this and the oxygen electrode.   TBM and RBM are in early stages of study and present fewer examples when compared to other biosensor membranes.   However, in line with the progress in biotechnology, the possibility of the production of a tissue-oriented sensor would increase once the structures and functions of animal and plant tissues have been better clarified.   In addition, a peptide hormone receptor has been isolated as a cell membrane fragment, and its structure is being determined through the analysis of genes.   In line with the progress of molecular biology concerning the receptor, the sensor which could utilize it may find a new application.

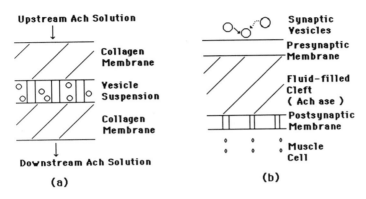

Upstream Ach Solution

Collagen Membrane

Vesicle Suspension

Collagen Membrane

Downstream Ach Solution

(a)

Synaptic Vesicles

Presynaptic Membrane

Fluid-filled Cleft (Ach ase)

Postsynaptic Membrane

Muscle Cell

(b)

Figure 4.8   Schematic diagram of the receptor biosensor membrane (AChI receptor) (a), and postsynaptic membrane (b).

SEMICONDUCTOR BIOSENSOR MEMBRANES (SBM)

The biosensor membranes discussed so far cover the cases which use an electrode as an electrochemical device.   Recently, an SBM consisting of ISFET (ion selective field effect transistor) and an EBM have been developed to take the place of electrodes, for the sake of high sensitivity and the compactness of the biosensor (Miyahara et al., 1983; Caras and Janata, 1980; Karube, 1985).   Owing to higher impedance, compactness is difficult for a glass electrode.   ISFET, in which an insulated gate-type field-effect transistor and ion sensor have been integrated, is compact (3mm max.) and has a lower output impedance. Figure 4.9 illustrates the structural diagram of a semiconductor biosensor membrane. SBM is cast on the gate insulation membranes (Si_3N_4/SiO_2) of the ISFET.   Because of the enzyme reaction, the ion concentration near the SBM varies and the drain current is modulated by the change in the interfacial electrical potential of Si_3N_4.   In this way the substrate corresponding to the modulation quantity or the concentration of product can be detected.

Since the semiconductor is highly sensitive and compact, the method of manufacturing the SBM requires a more refined technique with better reproducibility; e.g., an ultrathin membrane technology, with control of membrane thickness uniformity, semiconductor monolithic molding technology, integration technology, immobilization technology, etc. If future progress is made, there will be improvement in the response properties, life and reproducibility of semiconductor biosensors. This will gradually lead to industrial adoption.

Furthermore, there is a need for the development of ultrasensitive and ultracompact (10μm max.) microsensors (Castner and Wingard, 1984; Joseph, 1985; Simpson and Kobos, 1984). Microsensors will help make possible the elucidation of brain physiology, the direct detection of neurotransmitter, etc. In this connection, an ultrafine immobilizing technology would be indispensable.

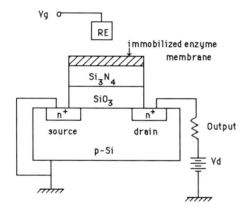

Figure 4.9 Example of a semiconductor biosensor membrane.

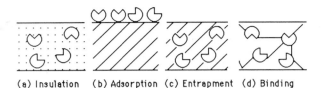

(a) Insulation   (b) Adsorption   (c) Entrapment   (d) Binding

Figure 4.10   Typical structures of biosensor membranes.

4.3 BIOSENSOR STRUCTURE AND THE METHOD OF IMMOBILIZATION

As stated earlier, biosensor membranes are formed by immobilizing molecular- identification elements onto synthetic or natural polymers. The differences in the structures of biosensor membranes under the various kinds of immobilizing methods are illustrated schematically in Figure 4.10.

INSULATION-BM

A suspension of enzymes, bacteria, mammalian tissue or plant tissue is sealed between gas-permeable membranes and UF membranes to form biosensor membranes. There are two cases extant for insulation-BM; in one case supports are not used, while in the other, supports of porous polymer membranes, etc., are used to adjust the quantity to be immobilized and overall membrane thickness. The latter type is called an impregnated biosensor membrane. This insulation-BM approach is applicable to all of the molecular identification elements available and particularly, it is effective with bacteria, mammalian and plant tissues, receptors, etc. It is used extensively in fundamental studies and it has been commercialized as a BOD sensor.

ADSORPTION-BM

Enzymes, antibodies and cells are adsorbed onto polymer membranes hydrophobically to form the adsorption-BM. The adsorption interaction varies substantially, depending upon the polymeric materials and their surface conditions. Polypropylene (Rice and Gold, 1984), onto which a protein adsorbs less readily than it does onto glass, has been used in making microtubes for biochemical experiments. Since nylon adsorbs protein well, porous nylon membranes have been developed as a blotting material for cell translation in genetic engineering. Since polycarbonate adsorbs cells, it has been used as a

Table 4.1 Adsorption of Various Substrates onto a Wet PVC Membrane

Substrates	M	Solvent	pH (Solvent)	N-Content (%)	Amount Sorbed ($\mu g/cm^2$)	Assay Method
Urea	60	H_2O	7.0	46.7	0	a
Glycine	75	H_2O	7.0	18.7	0	a
Uric acid	168	H_2O	7.0	33.3	0	a
Glucose	180	H_2O	7.0	--	0	c
Gly-gly-glycine	189	H_2O	7.0	21.7	0	a
Acetylcholine iodide	273	H_2O	7.0	--	0	b
Tetrahydro-cortisone	364	H_2O	7.0	--	0	d
Albumin	45000	H_2O	7.0	15.6	64.5	a
Hemoglobin	68000	H_2O	7.0	14.6	101	a
γ-globulin	156000	H_2O	7.0	14.7	122	a
Fibrinogen	400000	H_2O	7.0	10.6	198	a
Uricase	120000	borate	8.5	14.6	75.7	a
Glucose oxidase	186000	phosphate	5.6	13.0	69.9	a
β-glucuronidase	28000	acetate	7.0	14.4	80.9	a
Urease	480000	phosphate	7.0	12.7	88.6	a

a Micro-Kjeldahl method.
b Ion-electrode method described in this paper.
c Refer to Hirose et al., 1979.
d Refer to Hirose et al., 1980.

cell selecting plate for production of monoclonal antibodies.   Recently, enzyme-immobilizing carriers (membranes) have been developed in relation to a special process for preparing polyvinyl chloride (PVC) with an added adsorption function (Hirose, Yasukawa, Hayashi and Vieth, 1982) (Table 4.1).   The characteristics of these "wet" PVC membranes are that several kinds of enzymes can be immobilized simultaneously under normal conditions.   In the diagnosis of liver function, it is necessary to examine both glutamate pyruvate transaminase (GPT) and glutamate oxalacetate transaminase (GOT) rapidly.   It has been demonstrated that oxalacetate decarboxylase and pyruvate oxidase can be immobilized on "wet" PVC membranes simultaneously and that GOT/GPT can be measured consecutively using these membranes (Kihara et al., 1984) (see Figures 4.11 and 4.12). Many points remain to be clarified concerning polymeric materials and the hydrophobic mechanism of protein adsorption.   Studies are currently underway which are intended to clarify this function, and the development of adsorption-BM with better selectivity is anticipated in the future.

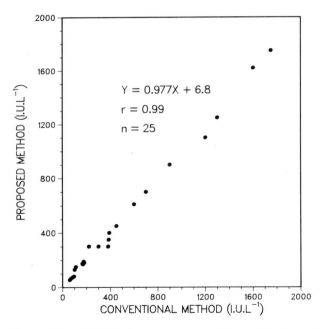

Figure 4.11   Relation between a spectrophotometric (conventional) method and the bienzyme biosensor (proposed) method in the determination of GOT activity.

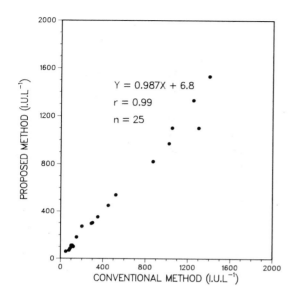

Figure 4.12   Relation between a spectrophotometric method and the bienzyme biosensor method in the determination of GPT activity.

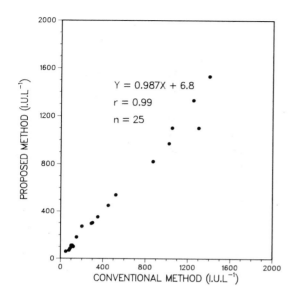

Figure 4.13   Structure of polyacrylamide-gel entrapped bacteria.

ENTRAPMENT-BM

Enzymes and bacteria are entrapped by - and fixed onto - polymer matrices to form the entrapment-BM. Enzymes and bacteria are entrapped into a three-dimensional macromolecular structure and this renders it difficult for them to elute from the matrix. On the other hand, substrates and products permeate and diffuse freely in the matrix. But, obviously, for macromolecular substrates, the entrapment-BM is unsuitable. A typical example of an entrapment-BM is a polyacrylamide gel and acrylic acid derivative (Figure 4.13). An enzyme is added to an aqueous solution of acrylamide to which the crosslinking reagents, N-N'-methylene-bis-acrylamide are added and the acrylamide is thus polymerized and cast onto a glass plate to form a thin layer. At this time, it is important to control the temperature so as not to deactivate the enzyme by an exothermic reaction. In contrast, since an entrapment-BM obtainable from the natural polymers, gelatin and carageenan, can be formed under normal, physiological-like, conditions, very little of the enzyme is deactivated. However, as the diffusion of substrate in the entrapment-BM can be slow, a problem is presented concerning the response properties of biosensor membranes.

BINDING-BM

Through the utilization of ions or functional groups present in the polymer matrix, an enzyme may be ionically bound, covalently bound and/or bonded by crosslinkage to form the binding-BM (Mascini, 1984). Lactase immobilization on DEAE-cellulose membranes is an example of the ionically-bound BM.

Enzymes may be covalently bonded to a matrix by the peptide method, the alkylation method, the diazo method, etc., to form covalently-bound biosensor membranes which experience little or no elution of enzyme and are, in that sense, more stable. Biosensors with better response properties were obtained by formulating thin polymer membranes this way. On the other hand, the formulation method is complicated and the enzyme tends to be deactivated chemically. Miyahara et al. (1983) utilize covalently-bound biosensor membranes for application as SBM. In order to attach the amino-group to the surface of the gate insulator Si_3N_4, the surface was first treated with (3-aminopropyl) triathoxysilane (APTES). This treatment prevented the stripping off of thin polymer membranes. Then, a thin membrane of triacetylcellulose was placed upon the FET by the dip method, and after treating the surface of triacetylcellulose membranes with glutaraldehyde, the FET was immersed into urease solution, and a urease-SBM was

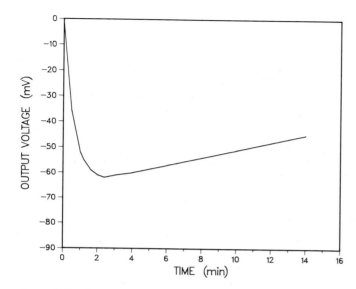

Figure 4.14   Time response of the differential mode enzyme FET (semiconductor BM) to 1 ml of 5×10^{-2} g/ml urea solution.   Conditions: 35.0°C, pH 7.0 and in 0.01 mol/dm-3 phosphate buffer solution.

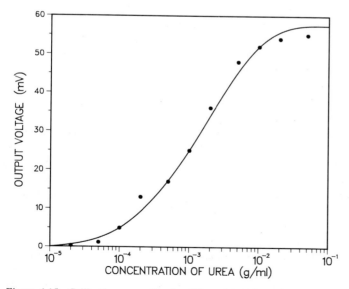

Figure 4.15   Calibration curve for the differential mode enzyme FET.   Conditions: 34.9°C, pH 7.0 at 1 min after urea solutions were poured in a 0.01 mol/dm-3 phosphate buffer solution.

obtained whose response characteristics are illustrated in Figs. 4.14 and 4.15.

Table 4.2   Performance of Biosensor Membranes

Fundamental studies with model solutions (reagents, control serum, control cultures, etc.)	Practical studies with actual solutions (blood of diseased patient, actual cultures, sewage water, etc.)
* Strength of biosensor membrane * Membrane formation 　　(immobilization, reproducibility) * Yield of immobilized enzyme (amt.) * Immobilized enzyme activity * Diffusion coefficient * Response time * Range and limit of detected 　concentration * Activity-profile with respect to 　pH, ionic strength, temp., buffers * Kinetic properties (K_m, V_{max}) * Specificity of substrate * Effects of interfering substances * Storage stability * Life time (half life of activity) * Reproducibility of output * Correlation with other standard 　methods	* Confirmation of fundamental 　properties using actual solutions * Comparison between batch and 　continuous systems * Comparison between rate assay 　and steady-state assay * Adsorption of interfering sub- 　stances on biosensor membranes * Injection of samples and buffers * Washing of biosensor membranes * Stirring in a reaction vessel * Control of temperature * Automation of apparatus * Standardized biosensor mem- 　branes (wet/dry, kit/cartridge, 　packing, guarantee)

PERFORMANCE CHARACTERISTICS OF BIOSENSOR MEMBRANES

The performance characteristics of biosensor membranes can be conveniently divided into two categories, one bearing on the fundamental studies of biosensor membranes using a model solution and the other for practical studies using an actual solution. The comparison appears in Table 4.2.

4.4 INTRODUCTION TO RESPONSE CHARACTERISTICS

The response characteristics of biosensor membranes represent the conditions of change in output during the period from the time when the specimen is added until the time when the output reverts to the original condition (Arnold and Rechnitz, 1984; Bergel and Comtat, 1984; Bradley and Rechnitz, 1984).   During a series of these intervals, the response properties of the biosensor correlate mutually with a variety of properties as shown in Table 4.2.   But a predominant issue one must consider is penetrant diffusion in biosensor membranes.   In general, Fick's law is applying,

$$\frac{\partial C}{\partial t} = D \frac{\partial^2 C}{\partial x^2} \qquad [4.1]$$

In order to experimentally measure the diffusivity in biosensor membranes, the time lag procedure is conveniently adapted.   Here, the relation between the diffused quantity (Q_t) and the time (t) is derived in the following formula:

$$Q_t = \int_0^t JAdt = \int_0^t [V \frac{dc}{dt} + qc] \, dt \qquad [4.2]$$

where J is the diffusive flux, A is the area of the biosensor membrane, V is the cell volume and q is sweep-fluid flow rate.   From the usual graphical relationship the time lag (θ) is estimated and the diffusion coefficient is obtained from the following equation:

$$D = \frac{L^2}{6\theta} \qquad [4.3]$$

where L is the membrane thickness of the BM.
    A diagram of the apparatus for measuring the diffusion coefficient in biosensor membranes is shown in Fig. 4.16, while a time lag plot for acetylcholine iodide (AChI) in a wet PVC-BM is shown in Fig. 4.17.   The diffusivity, D, varies with the material properties of the biosensor membrane.   It does not normally vary with the membrane thickness in homogeneous BM but does occasionally vary with the concentration of substrate.   A variation of D with the concentration of substrate is not desirable since the linear relationship between the sensor output and the concentration of substrate will not apply even in the range of relatively low concentration.

A typical biosensor membrane construction usually consists of de-proteinizing membranes (UF membranes) which remove impurities present in the solution and gas-permeable membranes, in addition to bioactive membranes made of fixed enzyme. Accordingly, in a practical biosensor, D' for a multilayer membrane construction must be taken into consideration. Barrie et al. (1963) have calculated D' in triplelayer membranes. Measurement of D' in a triple layer membrane system in which an RBM has been laminated between collagen membranes has been carried out in the Rutgers laboratory (Hirose and Vieth, 1984). In cases where the diffusivity is constant, the decrease in membrane thickness reduces the time lag (Θ) through a second power dependency and, thus, the sensor response time is sharply reduced. For this reason, thinning of the membranes of the BM is one of the most important subjects for evaluation. Thickness uniformity is evaluated, together with the strength of the biosensor membrane and its reproducibility in manufacturing.

The substrate concentration is another of the elements which influence the response properties. In the case of a high substrate concentration, both the rate of enzyme reaction and the rate of penetration of substrate into the biosensor membrane are usually accelerated, and the output of the sensor rapidly becomes linearly

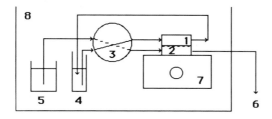

Figure 4.16   Apparatus for measuring diffusivities of a biosensor membrane: 1) biosensor membrane;   2) diffusion cell;   3) Masterflex pump;   4) recycled raidoactive ACh solution (10ml) (upstream);   5) flux buffer (pH 7.5);   6) microfuges (1.5 ml) for sampling;   7) stirrer;   8) water bath (3°C).

proportional to the concentration of substrate. Therefore, the range of concentrations to be detected, the detection limits (upper and lower), and the dilution method of the specimen solution need to be carefully evaluated.

As mentioned, the enzyme reaction condition is also one of the important elements which influence the response characteristics. The

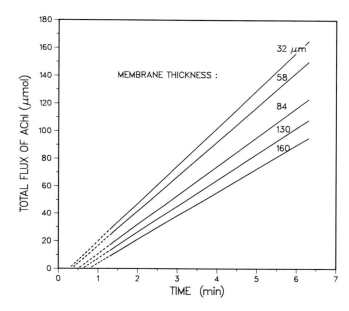

Figure 4.17 Plot of total flux of AChI against time under the conditions of various thicknesses of wet PVC membranes (for biosensor membranes).

response qualities of biosensors will be improved by increasing the activity of the enzyme and conducting the reaction under optimal conditions of pH, ionic strength, temperature, etc. The optimal conditions for a biosensor membrane often differ from those of the liquefied enzymes, due to environmental factors arising when enzymes are immobilized.   In addition, the possible elution and deactivation of enzymes present in biosensor membranes and the adsorption of interfering substances onto BM substantially influence the response properties.

As can be seen from the above, there are many points to be taken into consideration when selecting biosensor membrane material: to display a larger diffusivity; in simplifying a triple layer membrane to a single layer membrane (or thinning of multilayer membranes); in selecting an efficient immobilizing method, etc.

THE LIFE OF BIOSENSOR MEMBRANES

Enzymes will be deactivated by thermal and pH variations. Enzymes present in solution will usually deactivate within 2-3 days. Thus, one resorts to immobilization of an enzyme onto a polymer matrix to make the activity of the enzyme more lasting. For instance, glucose biosensor membrane onto which glucose oxidase has been immobilized has maintained its life for more than a month and has been commercialized. As such, the life of the biosensor membrane is one of the most important elements to be considered when producing a biosensor for practical use.

The measurement of BM life is conducted over a long test period of more than a month; the membrane is consecutively tested several thousand times. In cases where it is difficult to determine the life of biosensor membranes under the same conditions over a long period of time, it can be estimated by calculating the half-life of the enzyme activity. When it is desirable to determine the BM life in a short period of time, the time-course of activity is measured under several temperature conditions (e.g., 25-60° C). By extrapolation, the BM life can be swiftly estimated at normal temperature by this accelerated test. In the above experiments, it is often observed that biosensor membrane life will substantially differ in cases where a model solution is used, in comparison to the case where the practical solution for assay is used. Therefore, it is necessary to make a further detailed evaluation of the influence of inhibitors on the biosensor, the difference in the washing conditions of the membrane between usages, etc.

To build-in biosensor membrane life, it is desirable to measure it under the most suitable conditions for enzyme reaction. Recently, multi-BM made by immobilizing more than two kinds of enzymes are being studied (Rishpon, 1985; Scheller et al., 1985). Since suitable conditions for each enzyme vary, the optimum condition of the whole measuring system must be established.

Biosensor membrane life varies substantially, depending upon the preservation condition for the BM. Improvement in the length of the membrane life can be expected if it can be preserved in a dry condition rather than a wet one. Dry treatment technology together with surface treatment technology for the prevention of contamination should become important features in the development of functional polymeric materials.

4.5 APPLICATION CHARACTERISTICS OF BIOSENSOR MEMBRANES

The attributes of biosensor membranes and sensors in which BM are used are as follows:

(1)   Otherwise unstable molecular identification elements can be used repeatedly.

(2)   Specimens can be analyzed directly.

(3)   The analytical operation is simple.

(4)   Only a trace amount of the specimen is required.

(5)   No reagent other than a buffer solution is required.

(6)   Specimen clarity is not required.

(7)   Automatic measurement is enabled.

Examples of biosensor membranes in practice are shown in Table 4.3. Until now the most frequent use is in glucose measurement. The glucose-BM is being used for clinical examination, as well as in fermentation, food processing and medical treatment. In addition, there are the alcohol-BM, the lactate-BM, the BOD-BM, etc., available in practice. Below, two or three typical examples are described in more detail.

In 1970, an immobilized enzyme type of glucose-BM made by the Yellow Springs Instrument Company was commercialized. Since then, the glucose-BM has been developed by several companies in Japan. The various substances assayed in the automatic analysis of blood during clinical examination are shown in Table 4.4. Glucose is one of the important items in screening examinations as it is a basic product of metabolism in the human body. Since glucose examination requires the pre-treatment of blood, it does not permit measurement simultaneously with the other measuring items (multi; Table 4.4) and therefore, glucose (single; Table 4.4) has to be measured separately. The glucose biosensor method has the following merits, when compared to the colorimetric measurement method (HK-G6 PDH Method): firstly, the unit is compact and measurement is simple; secondly, BM life is long and almost maintenance-free; thirdly, a reagent is not required. Typical examples of glucose-BM are illustrated in Table 4.5.

Recently, a biosensor membrane capable of measuring three kinds of functions within a single layer has been developed. This has made it possible to measure one test specimen within 10 seconds and to assay 1500 test specimens per membrane. In the future, improved biosensor membranes for urea, creatinine, CPK, amylase, etc., are anticipated.

Table 4.3   Industrial Practice with Biosensor Membranes for Analysis

Biosensor Membrane (BM)	Manufacturer	Use	Assay item
Enzyme-BM	TOA Electric Co. Analytical Instr. Co. Fuji Electric Co. Yellow Springs Instr. Toyojozo Co.	Clinical Analysis Fermentation	Glucose Ethanol Lactic acid
	Ishikawa Co. Oriental Electric	Fermentation Food	Glucose
	Technicon Co. Kyoto Daiichi Elec. Co. Mitsubishi Chemical Co.	Clinical Analysis	Glucose
	Tateishii Electric Co. Nikkiso Fujisawa Pharmaceut.	Medical	Glucose Lactic acid
Bacterial BM	Denki Kagaku Keiki Nisshin Electric	Fermentation Environmental	Ethanol Acetic acid Biological oxygen demand
Semi-conductor BM	Mitsubishi Electric Co. Nippon Electric Co.		

Table 4.4   Clinical Assay and Autoanalyzer

Items	Biological Assay for Diagnosis				Autoanalyzer	
	Screening	Liver	Kidney	Thyroid	Multi	Single
Total protein	X	X	X	X	X	
Albumin	X	X	X	X	X	
A/G ratio[1]	X	X	X	X	X	
Total chole.[2]	X	X			X	
Glucose	X				X	X
N-urea	X		X	X	X	
Creatinine			X	X		
Uric acid			X		X	
Sodium	X		X	X	X	
Potassium			X	X		X
Chloride			X	X		X
Phosphorus	X		X	X	X	
Calcium	X		X	X	X	
GOT[3]	X	X			X	
GPT[4]	X	X			X	
LDH[5]	X	X			X	

[1] albumin/globulin;   [2] cholesterol;   [3] glutamate oxalacetate transaminase; [4]glutamate pyruvate transaminase;   [5] lactate dehydrogenase.

In the fermented food field, the concentration of glucose to be measured is high and thus, it presents dual problems;   i.e., the sampling of a solution in a fermentation tank and the method of its dilution, as well as the problem of bacterial removal.   Recently, a glucose-BM for fermentation that is capable of solving these problems has been developed.   A scheme of utilization of the glucose-BM for the automatic control of a fermentation system involving glucose, together with soluble oxygen, pH and alcohol, etc., is expected to be developed.

Measurement of BOD (biological oxygen demand) in waste water in the Warburg apparatus ordinarily takes about a week, thus making it desirable to develop a method of measuring it rapidly.   The compact BOD biosensor is characterized by its ability to measure BOD in a much shorter period of time.   BOD biosensor membranes available on the market are prepared by immobilizing the bacteria present in the sludge by the insulation method.   Materials for manufacturing membranes are available in kit form and can be used as teaching materials for biosensor studies.

Table 4.5   Examples of Commercialized Glucose Biosensor Membranes

Function of Biosensor Membrane	Polymer Materials			Comments
	"A Co."	"B Co."	"C Co."	
Gas-selective membrane	cellulose acetate	god -cellulose	wet -PVC[2]	permeated H_2O_2
Immobilized enzyme membrane	god [1] -cellulose			high activity of sensor membrane, rejection of protein, blood, corpuscles, etc.
Microfilter	polycarbonate	cellulose acetate		
Multilayer Membranes	3 layers	2 layers	1 layer	

[1] god: glucose oxidase;   [2] PVC: polyvinylchloride.

There are some biosensor membranes available as measuring systems for special factory use, though they are not broadly available on the market. For example, they are used to monitor amino acids, organic acids, etc., as well as utilization as unit monitors at the Aqua-Renaissance Project (Japan government).

Biosensor membranes which have evolved from the EBM owe much to the immobilizing technology which permits the utilization of a variety of molecular identification elements. Today, the state of development of electrochemical devices presents new opportunities to develop the SBM. Compact designs of the wire electrode and solid state electrode are progressing in electrochemical devices and it will not be long before micro-BM wields it power with finer materials and micromaterials (single cells, for example). On the other hand, in the molecular identification element field, an element with superior specific properties such as the monoclonal antibody is being developed, and the development of biosensor membranes which utilize this may take place. In line with the progress of electronics, the biosensor is being systemized, and a tactile biosensor and an image biosensor may someday be utilized.   Advanced BM may be multifunctional and integrated, as on a biochip (Karube, 1985) and a biocomputer element (Amari, 1985).   As has been observed with regard to the development of biosensor membranes, the significance of the parallel development of functional polymeric materials in biochemistry and electrochemistry is growing daily.

This chapter goes on now to a consideration of systems analysis in somewhat more detail.

4.6 EFFECT OF A FIRST ORDER REACTION ON THE PENETRANT TIME LAG IN A MEMBRANE BIOSENSOR

Ludolph et al. (1979) derived an expression for the time lag in a system influenced by the effects of a linear irreversible reaction.   The transient equation becomes:

$$\frac{\partial C}{\partial t} = D \frac{\partial^2 C}{\partial x^2} - kC \qquad [4.4]$$

with $C(x,0) = 0$, $C(0,t) = C_0$ and $C(l, t) = 0$ as the boundary conditions.

By manipulating equation [4.4] and its steady state solution, $C_s(x)$, it is possible to obtain an expression for the time lag, Θ, without solving the full transient problem.   The expression is:

$$\Theta = \frac{\int_0^l C_s \, dx + k \int_0^l u[x]dx - \frac{1}{C_0} \int_0^l C_s^2 \, dx}{J_s [l]} \qquad [4.5]$$

where $u(x)$ satisfies the boundary value problem:

$$D \frac{d^2 u}{dx^2} - ku = C_s \qquad [4.6]$$

$$u[0] = u[l] = 0 \qquad [4.7]$$

Upon integration, a rather complex expression results:

$$\frac{\Theta D}{\mathsf{L}^2} = f [R \, \mathsf{L}] \tag{4.8}$$

where $R = (k/D)^{1/2}$. It may be simplified to the following form:

$$\frac{\Theta D}{\mathsf{L}^2} = \frac{2R \, \mathsf{L} \cosh [R \, \mathsf{L}] - 2 \sinh [R \, \mathsf{L}]}{4R^2 \, \mathsf{L}^2 \sinh [R \, \mathsf{L}]} \tag{4.9}$$

Leypoldt and Gough (1980) presented an elegant solution for the concentration field by using finite Fourier transforms, leading to the useful expression:

$$\frac{\Theta D}{\mathsf{L}^2} = \frac{1}{2} [\coth \phi / \phi - 1 / \phi^2] \tag{4.10}$$

where $\phi^2 = \dfrac{k\mathsf{L}^2}{D} = R^2 \mathsf{L}^2$

Inspection shows that equation [4.9] readily reduces to this form.

Leypoldt and Gough further showed that, in the case of reversible penetrant immobilization where $D_{eff} = D/(\mathcal{R} + 1)$, equation [4.10] becomes:

$$\frac{\Theta D}{[\mathcal{R} + 1] \, \mathsf{L}^2} = \frac{1}{2} [\coth \phi / \phi - 1 / \phi^2] \tag{4.11}$$

In the limit of no reaction the expression reduces to the familiar one,

$$\Theta = \frac{\mathsf{L}^2}{6D} [\mathcal{R} + 1] \tag{4.12}$$

The results are presented in Figure 4.18.

The main conclusion is that the effect of an irreversible first-order reaction is to *decrease the time lag*. This is in contrast to the

effect of equilibrium reversible penetrant immobilization which is to
increase Θ.

4.7 SPATIAL DISTRIBUTION OF BIOCATALYST

In the case of a reaction in a membrane matrix, transport rates
and reaction rates of substrates and products are, of course,
interrelated.   Reaction can speed up the transport (facilitation) or
transport can slow down the reaction rate.   When several of these
processes are coupled or connected in the proper vectorial way, the

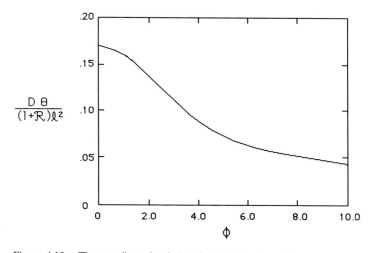

Figure 4.18   The non-dimensional time lag as a function of ϕ.

phenomenon is termed 'chemiosmotic energy conversion.'   According to
the Curie Principle, a symmetric enzyme distribution system in a
membrane operating with symmetric boundary conditions cannot create
any directional difference of flux of any reactant (deGroot and Mazur,
1962).   Therefore, to direct a steady flux of a reactant under symmetric
boundary conditions, the membrane must be in some way vectorized.
Thus, when an asymmetric membrane is used, some part of the actual
driving force (affinity for the proceeding chemical reaction) is needed
for maintaining a nonzero difference of the two diffusion fluxes of the
product (i.e., fluxes into substrate and product chambers) (see Fig.
4.19).   Actually, the asymmetry needed could also reside in the spatial

variation of the diffusivity of the penetrant in the matrix instead of the enzymatic activity.

Kubin and Spac´ek (1973) theoretically showed that under symmetrical boundary conditions, utilizing an enzymatically asymmetric membrane, it is possible to direct higher flux of the product toward one side of the membrane. Blumenthal et al. (1967) also showed that, using an asymmetric membrane, it is possible to couple the reaction to the flow of solutes or to the flow of electrical currents. They used two ion-exchange membranes of opposite charges separated by an intermediate layer of solution containing an enzyme. Another model system studied is a bilayer membrane with two different enzymes for consecutive reactions (Broun et al., 1972).

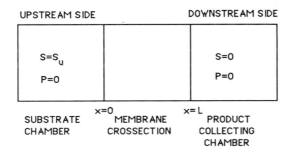

Figure 4.19   The reactor system and the boundary conditions.

A different, but perhaps more realistic, situation arises when asymmetric boundary conditions are employed. For experimental and/or simulation purposes, the case of a diffusion cell arrangement is convenient where the following are the appropriate boundary conditions (see Fig. 4.19):

$$
\begin{array}{lll}
x = 0 & S = S_u & P = 0 \\
x = L & S = 0 & P = 0
\end{array}
$$

Goldman et al. (1968) have performed some experimental studies with such a system. They showed that backflow of product into the substrate reservoir occurs at a higher rate than the flow of product into the product collecting chamber. Clearly, this is an undesirable situation, especially when the product flux is to be monitored, as in enzyme membrane sensor applications.

Çiftçi and Vieth (1980) looked into improvement of directed product flux, sometimes at the expense of lower fractional conversion of substrate to product.

To begin with, consider a reversible enzyme reaction with the immobilized enzyme. For the reaction,

$$E + S \underset{k_2}{\overset{k_1}{\Leftrightarrow}} E \cdot S \underset{k_4}{\overset{k_3}{\Leftrightarrow}} P + E \qquad [4.13]$$

coupled with the diffusion in the membrane, one can develop the following steady state equations:

$$D_s \frac{d^2 S_x}{dx^2} = V_m \cdot \frac{S_x - a \cdot P_x}{k_m + S_x + bP_x} \qquad [4.14]$$

and

$$D_p \frac{d^2 P_x}{dx^2} = - V_m \cdot \frac{S_x - a \cdot P_x}{k_m + S_x + bP_x} \qquad [4.15]$$

with boundary conditions:
$$x = 0 \qquad S = S_u \qquad P = 0$$
$$x = L \qquad S = 0 \qquad P = 0$$

where D_s, D_p are diffusion coefficients of substrate and product in the membrane. S_x, P_x are concentrations of substrate and product at position x in the membrane.

$$V_m = k_3 \cdot E_T \qquad = V_m^{S \to P} \qquad [4.16]$$

$$k_m = [k_2 + k_3] / k_1 \qquad = k_m^S \qquad [4.17]$$

$$a = k_2 \cdot k_4 / k_1 \cdot k_3 \qquad = \frac{V_m^{P \to S} k_m^S}{V_m^{S \to P} \cdot k_m^P} \qquad [4.18]$$

$$b = k_4 / k_1 \qquad = \frac{k_m^S}{k_m^P} \qquad [4.19]$$

This model can be simulated using numerical computation techniques to obtain steady state concentration profiles and fluxes for various cases with different coefficients (Çiftçi, 1978). For a general case (A) we obtain the flux ratios shown in Table 4.7 (see Table 4.6 for data).

Table 4.6   Data for Base Case (A)

$V_m^{S \to P}$ = 1.512 μM/(cc swollen membrane x s)
K_m^S = 407.5 μM/(cc swollen membrane)
$V_m^{P \to S}$ = 0.177 μM/(cc swollen membrane x s)
K_m^P = 624.3 μM/(cc swollen membrane)
D_s = 0.57 x 10^{-7} cm^2/s
D_p = 0.57 x 10^{-7} cm^2/s
S_u / K_m^S = 4 (dimensionless)
L = 0.0173 cm

Table   4.7   Dimensionless Fluxes for Base Case (A)

$\dfrac{F_{pd}}{F_{su}}$	$\dfrac{F_{sd}}{F_{su}}$	$-\dfrac{F_{pu}}{F_{su}}$	$\dfrac{F_{su}}{F_{snr}}$
0.252	0.032	0.717	3.53

Flux ratios have been presented for the sake of simplicity. F_{pd} is product flux into the product collecting chamber; F_{su} is substrate flux into the membrane from the substrate chamber; F_{sd} is substrate (unconverted) flux into the product collecting chamber; $-F_{pu}$ is the product flux into the substrate chamber, and F_{snr} is substrate flux into and out of the membrane in the absence of reaction. F_{pd}/F_{su} gives the fraction of product collected in the product chamber per unit amount of substrate diffusing into the membrane. F_{sd}/F_{su} shows the fraction of substrate transported without reaction into the product collecting chamber. $-F_{pu}/F_{su}$ shows the fraction of backflux of product into the substrate chamber for unit amount of substrate diffusing into the membrane. F_{su}/F_{snr} shows the *facilitation* of substrate flux into the membrane because of reaction compared with the case of no reaction. These results are in good agreement with the results of experimental work mentioned above (Goldman et al., 1968).   A detailed explanation of the calculational procedures of the simulation study by Çiftçi (1978) is given elsewhere.   Consideration of the behavior of anisotropic enzyme membrane structures was the next order of business.

4.8 ANISOTROPIC ENZYME DISTRIBUTION

A number of anisotropic cases with simple graphical representations are given in Table 4.8.   Notice the changes in the values of fractions of backflux and forward flux of product when the enzyme concentration is shifted toward the downstream side.

The simple conclusion which can be drawn from this analysis is that the location of the enzyme molecules controls the directionality of fluxes as well as flux magnitudes.   Clearly, the optimal distribution for maximizing emergent product flux will be one where the enzyme concentration is shifted toward the downstream side of the membrane. This will be of importance in enhancing the current response at an enzyme electrode (see Figure 4.20) or in strengthening the level of a synaptic response to an instantaneously maximal flux of chemical messenger, as will be discussed in Chapter 9.

Table   4.8   Effect   of   Different   Linear   Enzyme   Concentration Distributions   on   Dimensionless   Fluxes

Case	Enzyme concentration profile	F_{pd}/F_{su}	F_{sd}/F_{su}	$-F_{pu}/F_{su}$	F_{su}/F_{snr}
A		0.252	0.032	0.717	3.53
B		0.222	0.068	0.71	3.45
C		0.385	0.107	0.507	2.03

Enzyme activity varies linearly between 0.0 &0.25 μM/cc swollen membrane x s.

4.9 THEORETICAL MODEL FOR ANISOTROPIC ENZYME MEMBRANES

Pedersen and Chotani (1981) took the next logical step, examining the case where the enzyme is confined within a surface layer on a membrane support. Recently, The´venot et al. (1979) and Coulet et al. (1980) have developed enzyme electrodes that have an enzyme immobilized as a *surface* layer on a highly polymerized collagen film. In particular, immobilized glucose oxidase was employed for the conversion of glucose (analyte) to hydrogen peroxide that was subsequently detected by a platinum electrode. The system provides a sensitive and stable assay for glucose (Coulet and Gautheron, 1976) as well as a simple method for the investigation of the interplay of reaction and diffusion in a heterogeneous system (Coulet et al., 1980; Horvath and Engasser, 1974; Engasser and Horvath, 1973; Engasser et al., 1977).

The physical model and coordinate system for an anisotropic enzymic membrane are shown in Fig. 4.20. Reactant diffusing through the supporting membrane from the bulk solution is converted by an interfacial enzymatic reaction to a product that is subsequently detected by an appropriate sensor. For example, an amperometric peroxide-sensing electrode can be used to detect hydrogen peroxide formed from the glucose oxidase-catalyzed reaction involving glucose and oxygen (The´venot et al., 1979). The local reactant concentration at

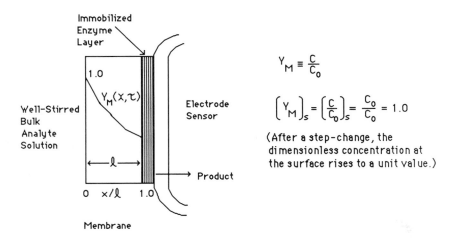

$$Y_M \equiv \frac{C}{C_0}$$

$$\left[Y_M\right]_s = \left[\frac{C}{C_0}\right]_s = \frac{C_0}{C_0} = 1.0$$

(After a step-change, the dimensionless concentration at the surface rises to a unit value.)

Figure 4.20   Membrane system (enzyme downstream).

the catalyst surface is so low that the reaction kinetics are adequately described by a first-order rate law. This latter assumption assures that the electrode response is proportional to the analyte concentration in the absence of nonlinearities in the transport phenomena, consistent with what is usually observed and desired in practice.

The governing dimensionless equation describing reactant transport within the membrane is:

$$\frac{\partial Y_M}{\partial \tau} = \frac{\partial^2 Y_M}{\partial X^2}$$ [4.20]

where the dimensionless time τ, penetration X and membrane substrate concentration Y_M variables are defined as:

$$\tau = Dt/l^2$$ [4.21a]

$$X = x/l$$ [4.21b]

$$Y_M = C/C_0$$ [4.21c]

The membrane thickness is l, and D is the diffusion coefficient. Equation [4.20] is used with the boundary and initial conditions:

$$Y_M = 1 \qquad\qquad \tau > 0 \qquad\qquad X = 0$$ [4.22a]

$$\frac{\partial Y_M}{\partial X} = -\mu Y_M \qquad\qquad \tau > 0 \qquad\qquad X = 1$$ [4.22b]

$$Y_M = 0 \qquad\qquad \tau = 0 \qquad\qquad 0 \leqslant X \leqslant 1$$ [4.22c]

Using standard techniques, the following expressions can be obtained,

$$Y_M(X) = 1 - \eta \mu X$$ [4.23a]

where $\eta = [1 + \mu]^{-1}$ and,

$$Y_M(X, \tau) = 1 - \eta \mu X + 2 \sum_{n=1}^{\infty} \frac{\sin \lambda_n X}{\cos \lambda_n \sin \lambda_n - \lambda_n} \exp[-\lambda_n^2 \tau]$$ [4.23b]

From its definition, the time lag is obtained as:

$$\frac{D\theta}{\ell^2} = 2\,[1+\mu]\sum_{n=1}^{\infty}\frac{\sin\lambda_n}{\lambda_n^2\,[\lambda_n - \cos\lambda_n\,\sin\lambda_n]}\qquad [4.24]$$

which can be approximated by the following expression with a high degree of accuracy:

$$\frac{D\theta}{\ell^2} = \frac{3+\mu}{6\,[1+\mu]}\qquad [4.25]$$

From equation [4.25] it is apparent that as μ increases, the time lag approaches the minimum value $\theta = \ell^2/6D$, while as μ decreases, the time lag increases to values approaching $\theta = \ell^2/2D$, in contradistinction to what is observed for isotropic membrane reactors as described earlier in this chapter.

The predicted electrode response is shown in Figures 4.21 and 4.22 as a function of the modulus μ. The values in Fig. 4.21 show the dimensionless product flux, from eqn. [4.20b], normalized to the steady state value, $\eta\mu$. In practice, the sensor output reading is followed until the steady state value is obtained. At that time, the electrode response is calibrated with the analyte concentration. As seen in Fig. 4.21, however, for low enzyme loadings, the time to reach a steady state value may be extremely large. Rather, it is more expeditious to follow the *derivative* of the sensor response and make use of the sensor reading at the peak value of the derivative; i.e., at the inflection point of the original response curve. The predicted values obtained by this derivative method are shown in Fig. 4.22 for the corresponding curves of Fig. 4.21. All the curves have been normalized to the maximum response value. The time required to carry out a single analysis is considerably reduced, at least fivefold. Furthermore, from the location of the peak heights, it is also seen in Fig. 4.22 that the time necessary to reach the response signal recording value is less sensitive to the values of the modulus, relative to what is observed in Fig. 4.21. This implies that the derivative method is not only a more rapid procedure for carrying out assays with enzyme electrodes than the direct response method, but is also less sensitive to loss of enzyme activity as regards the assay time.

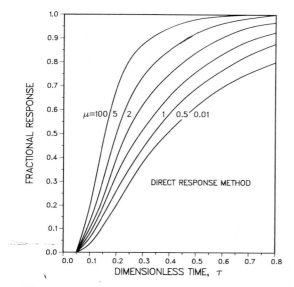

Figure 4.21   Electrode response as a function of the dimensionless time τ with the modulus μ as a parameter.   The direct response method is depicted for amperometric detection.

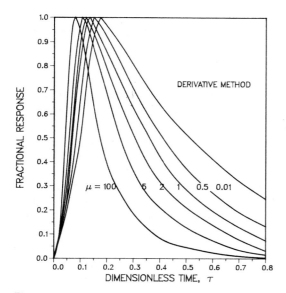

Figure 4.22   Electrode response measured by the derivative method as a function of the dimensionless time τ with the modulus μ as a parameter.   The curves correspond to the derivative function of the corresponding curves in Fig. 4.21 when normalized to the peak derivative value.

The derivative response method has been exploited recently in connection with an anisotropic membrane for a glucose sensing electrode (The´venot et al., 1979; Coulet et al., 1980). The experimental data obtained are in qualitative agreement with the curves shown as Figures 4.21 and 4.22. For example, membranes 0.3 mm thick with an analyte diffusion coefficient given as $2(10^{-6})$ cm^2/s would be expected to have a dynamic response time between 30 and 50 s, according to Fig. 4.22. This corresponds to the experimental values reported by The´venot et al. (1979).

4.10 REFERENCES

Amari, *Kobunshi*, **34**, 382 (1985).

Arnold, M.A. and G.A. Rechnitz, *Anal. Chim. Acta.*, **113**, 351 (1980).

Arnold, M.A. and G.A. Rechnitz, *Anal. Chem.*, **53**, 515 (1981).

Arnold, M.A. and G.A. Rechnitz, *Anal. Chem.*, **53**, 1837 (1981).

Arnold, M.A. and G.A. Rechnitz, *Anal. Chem.*, **54**, 777 (1982).

Arnold, M.A. and M.E. Meyerhoff, *Anal. Chem.*, **56**, 20R (1984).

Arnold, M.A. and G.A. Rechnitz, *Anal. Chim. Acta.*, **158**, 379 (1984).

Arnold, M.A., *Anal. Chem.*, **57**, 565 (1985).

Barrie, J.A., J.D. Levine, A.S. Michaels and P. Wong, *Trans. Faraday Soc.*, **59**, 869 (1963).

Bergel, A. and M. Comtat, *Anal. Chem.*, **56**, 2904 (1984).

Blum, L.J. and P.R. Coulet, *Anal. Chim. Acta.*, **161**, 355 (1984).

Blumenthal, R., S.R. Caplan and O. Kedem, *Biophys. J.*, **7**, 735 (1967).

Boitieux, J.L. and D. Thomas, *Anal. Chim. Acta.*, **163**, 309 (1984).

Bradley, C.R. and G.A. Rechnitz, *Anal. Chem.*, **56**, 664 (1984).

Broun, G., D. Thomas and E. Selegny, *J. Memb. Biol.*, **8**, 313 (1972).

Bush, D.L. and G.A. Rechnitz, *J. Memb. Sci.*, **30**, 313 (1987).

Caras, S. and J. Janata, *Anal. Chem.*, **52**, 1935 (1980).

Castner, J.F. and L.B. Wingard, *Anal. Chem.*, **56**, 2891 (1984).

Çiftçi, T., M.S. Thesis in Chemical and Biochemical Engineering, Rutgers U. (1978).

Çiftçi, T. and W.R. Vieth, *J. Mol. Catal.*, **8**, 455 (1980).

Coulet, P.R. and D.C. Gautheron, in "Analysis and Control of Immobilized Enzyme Systems," North-Holland: Amsterdam, p. 165 (1976).

Coulet, P.R., R. Sternberg and D.R. The´venot, *Biochim. Biophys. Acta.*, **612**, 317 (1980).

deGroot, S.R. and P. Mazur, "Non-equilibrium Thermodynamics," North Holland: Amsterdam, p. 57 (1962).

Engasser, J.-M. and C. Horvath, *J. Theor. Biol.*, **42**, 137 (1973).

Engasser, J.-M., P.R. Coulet and D.C. Gautheron, *J. Biol. Chem.*, **252**, 7919 (1977).

Goldman, R., O. Kedem and E. Katchalski, *Biochemistry*, **7**, 4518 (1968).

Guilbault, G.G., *Anal. Chem.*, **55**, 1682 (1983).

Haga, Itagaki and Okano, *Nihon Kagakukaishi*, 1549 (1980).

Hirose, S., E. Yasukawa, M. Hayashi and W.R. Vieth, *J. Memb. Sci.*, **11**, 177 (1982).

Hirose, S. and W.R. Vieth, *Appl. Biochem. Biotech.*, **9**, 81 (1984).

Horvath, C and J.-M. Engasser, *Biotechnol. Bioeng.*, **16**, 909 (1974).

Ikariyama, Y., M. Furuki and M. Aizawa, *Anal. Chem.*, **57**, 496 (1985.

Joseph, J.P., *Anal. Chim. Acta.*, **167**, 249 (1985).

Karube, I., *Yukagaku*, **32**, 95 (1983).

Karube, I., *Biochem. Eng. VII Abstracts*, Engineering Foundation Conferences, New York (1985).

Karube, I., *Kobunshi*, **34**, 386 (1985).

Keating, M.Y. and G.A. Rechnitz, *Anal. Chem.*, **56**, 801 (1984).

Kihara, K., E. Yasukawa and S. Hirose, *Anal. Chem.*, **56**, 1876 (1984).

Kubin, M. and P. Spac´ek, *Polymer*, **14**, 505 (1973).

Kuriyama, S. and G.A. Rechnitz, *Anal. Chim. Acta.*, **131**, 91 (1981).

Kuriyama, S., M.A. Arnold and G.A. Rechnitz, *J. Memb. Sci.*, **12**, 269 (1983).

Leypoldt, J.K. and D.A.Gough, *J. Phys. Chem.*, **84**, 1058 (1980).

Ludolph, R.A., W.R. Vieth and H.L. Frisch, *J. Phys. Chem.*, **83**, 2795 (1979).

Mascini, M., D. Moscone and G. Palleschi, *Anal. Chim. Acta.*, **157**, 45 (1984).

Miyahara, Moriizumi, Shiokawa, Matsuoka, Karube and Suzuki, *Nihon Kagakukaishi*, 823 (1983).

Nikolelis, D.P., *Anal. Chim. Acta.*, **167**, 381 (1985).

Pau, C.P. and G.A. Rechnitz, *Anal. Chim. Acta.*, **160**, 141 (1984).

Pedersen, H. and G.K. Chotani, in "Applied Biochemistry and Biotechnology," **6**, Humana Press Inc., p. 309 (1981).

Rechnitz, G.A., *Science*, **214**, 287 (1984).

Rice, M.R. and H.S. Gold, *Anal. Chem.*, **56**, 1436 (1984).

Rishpon, J., *Biochem. Eng. VII Abstracts*, Engineering Foundation Conferences, New York (1985).

Scheller, F., R. Renneberg and F. Schubert, *Biochem. Eng. VII Abstracts*, Engineering Foundation Conferences, New York (1985).

Schubert, F., F. Scheller and D. Kirstein, *Anal. Chim. Acta.*, **141**, 15 (1982).

Schubert, F., R. Renneberg, F.W. Scheller and L. Kirstein, *Anal. Chem.*, **56**, 1677 (1984).

Schubert, F., D. Kirstein, K.L. Schroder and F.W. Scheller, *Anal. Chim. Acta.*, **169**, 391 (1985).

Seitz, R.W., *Anal. Chem.*, **56**, 16A (1984).

Simpson, D.L. and R. Kobos, *Anal. Chim. Acta*, **164**, 273 (1984).

Suzuki, S., Ed., "Ion Electrodes and Enzyme Electrodes," Kodansha 1981).

Suzuki, S., Ed., "Chemical Sensors," Kodansha (1983).

Suzuki, S., Ed., "Biosensors," Kodansha (1984).

The´venot, D.R., R. Sternberg, P.R. Coulet, J. Laurent and D.C. Gautheron, *Anal. Chem.*, **51**, 96 (1979).

Updike, S.J. and G.P. Hicks, *Nature*, **214**, 986 (1967).

Wohltjen, H., *Anal. Chem.*, **56**, 87A (1984).

Some days you eat the bear.
Some days the bear eats you.

Trad. Folk Saying

5

ENZYME AND CELL-BASED REACTORS

5.0 BIOREACTORS

Although only recently appearing on the scene, bioreactors have already found large scale industrial application in the food and pharmaceutical industries (Chibata, 1986) and a wide variety of smaller scale applications in medical diagnostics and therapeutics (Pedersen and Horvath, 1981; Bernath et al., 1976).

Chibata (1986) lists examples of current large scale processes as follows:

1. Production of L-amino acids from acetyl-DL-amino acids using immobilized aminoacylase.
2. Production of 6-aminopenicillanic acid from penicillin G using immobilized penicillin amidase.
3. Production of high fructose syrup using immobilized glucose isomerase.
4. Production of L-aspartic acid using immobilized microbial cells.
5. Production of L-malic acid using immobilized microbial cells.
6. Production of L-alanine using immobilized microbial cells.
7. Hydrolysis of lactose in milk using immobilized β-galactosidase.

In addition, studies on immobilization of animal and plant cells are actively underway (e.g., Karkare et al., 1986; Prenosil and Pedersen, 1983).

This spectrum of activity can be conveniently represented in a "bioconversion network," as shown in Fig. 5.1, which also indicates the interaction of classical fermentation technology and biotechnology.

Pedersen and Horvath (1981) point out that the most widespread use so far of immobilized enzymes in analytical applications (e.g., medical diagnostics) is the employment of enzyme tubes in continuous-flow analyzers of the Technicon type. Highly active enzymic layers are deposited on the inner walls of plastic tubes, which are coiled into modular forms of open tubular heterogeneous enzyme reactors (dubbed "others"). Under the conditions of operation, the reactors are diffusion controlled, displaying "linear chemistry;" i.e., the pseudo first order kinetic conditions assure chemical analytical linearity.

5.1 REACTORS WITH POROUS ANNULAR CATALYTIC WALLS

In the case of a tubular reactor with a porous annular catalytic shell at the tube wall, a solid-phase equation is, of course, necessary. The solid-phase equation, which equates the diffusive flux of substrate inside the porous annulus to the reaction rate of substrate conversion, incorporates a modified Thiele modulus, ϕ_m, which is based on the thickness of the porous catalytic membrane-like annulus rather than on the half-thickness of a catalyst pellet. An effectiveness factor, η, can be defined and evaluated for different substrate concentration values. Numerical solutions of these equations have been obtained by Horvath et al. (1973a). In order to simplify the computational algorithm, the effectiveness factor was expressed as a weighted sum of the effectiveness factors for zero- and first-order reactions (limiting orders of Michaelis-Menten kinetic expression).

Simulation results from this model are presented in Fig. 5.2. The regimes of diffusion control and kinetic control are highlighted in terms of a dimensionless parameter called reactor modulus. It is a ratio of the maximum possible rate in the catalytic annulus to the maximum possible rate of radial molecular diffusion. The reactor performance can be described in terms of the reactor modulus and dimensionless reactor length, ζ, which is proportional to the average rate of reaction to achieve a desired conversion level. As shown in Fig. 5.2, at reactor modulus values far greater than unity, the reaction is bulk diffusion-controlled; ζ is independent of reactor modulus in this region. At low modulus values, the reaction becomes kinetically controlled. These limiting functional dependencies of ζ for diffusion control and for

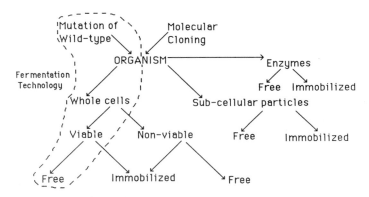

Figure 5.1 Bioconversion network.

kinetic control may be represented as ζ_{diff} and ζ_{kin} , respectively.   The intermediate reactor modulus values characterize the transition regime. The reactor length ζ required to achieve a particular conversion level can be calculated by the approximate relation:

$$\zeta \approx \zeta_{diff} + (\zeta_{kin} / \overline{\eta})$$ [5.1]

where $\overline{\eta}$ is an average catalyst effectiveness factor.   The second term on the right-hand side may be construed as ζ_{cat} , i.e., the axial distance required to achieve a desired conversion when the overall reaction is pore   diffusion-controlled.   This procedure   of evaluating ζ values by

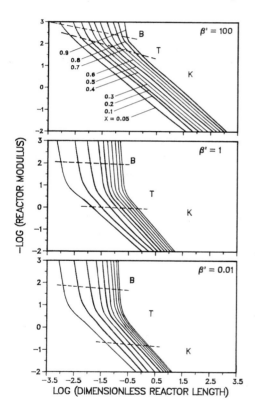

Figure 5.2   Relationship between reactor modulus and dimensionless reactor length required to achieve conversions indicated at different values of dimensionless substrate concentration, $\beta'(= S_O/K'_m)$, and fractional conversion, χ.   Note the regimes of bulk diffusion control (B), kinetic control (K), and the transition regime (T).   The physical system considered here is an enzyme reactor with porous annular walls (From Horvath et al., 1973b).

simple additivity of the limiting ζ values permits a simplified treatment of the reactor data and problems of reactor design.

Based on the above definitions of the reactor coordinate, Horvath et al. (1973b) have also proposed a reactor effectiveness factor. It is expressed as the ratio of the actual reaction rate to some maximum reaction rate characteristic for that reactor, both evaluated for a given conversion. Thus, the efficiency of the reactor can be represented by a kinetic effectiveness factor, E_{kin} , and a diffusional effectiveness factor, E_{diff}:

$$E_{kin} = \zeta_{kin} / \zeta \qquad [5.2]$$

$$E_{diff} = \zeta_{diff} / \zeta \qquad [5.3]$$

E_{kin} is a measure of the degree of utilization of the enzyme because it is a ratio of the actual heterogeneous catalytic rate to the rate that would be obtained if the same amount of enzyme were used in a homogeneous plug-flow reactor under the same conditions. The actual reaction rate relative to the rate that would result if sufficient enzyme were present at the catalytic wall such that the surface concentration of substrate would be essentially zero is given by E_{diff}. Plots of the type shown in Fig. 5.3 have been presented which are useful to evaluate the reactor effectiveness factors. From Fig. 5.3 it may be noted that the dependence of E_{kin} on the reactor modulus is similar to that of the catalyst effectiveness factor on the modified Thiele modulus. Using these theoretical developments, reactor design equations for the limiting cases of bulk diffusion-control and kinetic-control have been presented (Horvath and Solomon, 1972). Correlations of Sherwood number with dimensionless reactor lengths from the chemical engineering literature have been used in developing these design equations. The foregoing analysis has also been extended to shell-structured enzyme reactors - known as pellicular enzyme reactors - in which the catalyst is confined to a membrane-like porous outer shell surrounding an impervious inert core (Horvath and Engasser, 1973).

These remarks highlight once again the importance of membrane diffusion and reaction phenomena in governing the performance of a membrane appliance. A comparison of relative potencies of several reactor designs is described by Vieth et al. (1976). To provide additional background and perspective, it seems worthwhile to update and refine those findings here.

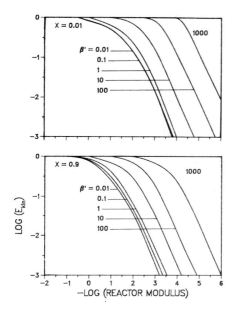

Figure 5.3   Reactor effectiveness factor, E_{kin}, as a function of reactor modulus for various β' values at 1% and 90% conversion levels. The reactor system is the same as that for Fig.5.2. (From Horvath et al., 1973b).

5.2 COMPARISON OF RELATIVE POTENCIES OF SEVERAL REACTOR DESIGNS

The overall potency of an enzyme reactor is determined by myriad factors, as highlighted in the foregoing discussion. Although some general analyses and design procedures for enzyme reactors are now known, the relative paucity of experimental information - particularly on a pilot-plant scale - renders difficult the task of designing and scaling up an enzyme reactor. A designer, contemplating a given enzymic reaction system, may be faced with several alternative approaches - perhaps none of them having a clear advantage or design precedence. If comparative information were available on different alternative design schemes, it would facilitate the choice of a particular system. Despite the large body of information available in the literature, a direct comparison of this type is not easy, owing to many variations in experimental and/or theoretical approaches. Presented below is a simplified attempt to compare the relative efficiencies of several reactor configurations, based on research done in our laboratory on free and immobilized enzymes. These data were obtained under conditions that permit their meaningful comparison.

The most important factors governing the overall reactor efficiency include the enzyme loading factor, carrier loading factor, operational stability of enzyme, external and internal diffusional efficiency, and residence time distribution. The efficacy of different reactor configurations with respect to these factors is compared in Table 5.1. Data on continuous enzyme reactors using free (Bowski et al., 1972), bead-immobilized (Saini et al., 1972), microencapsulated (Mogensen and Vieth, 1973) and collagen membrane-immobilized (Venkatasubramanian et al., 1972; Eskamani, 1972; Wang and Vieth, 1973; Constantinides et al., 1973; Bernath and Vieth, 1974) enzymes and collagen-immobilized whole microbial cells (Saini and Vieth, 1975) formed the basis for the comparison presented in Table 5.1.

These comparisons are made according to eqns. [5.7] and [5.8], which employ pseudo first-order kinetics. Equation [5.7] refers to flat sheets of collagen-enzyme membranes, and eqn. [5.8] relates to spherical beads or microcapsules.

FIRST-ORDER, IRREVERSIBLE REACTION

For a fixed-bed reactor containing an enzyme-membrane, Vieth et al. (1976) developed expressions to describe different reaction kinetic schemes. The physical system modeled is the spiral-wound multichannel reactor shown in Figure 3.17. Considering the membrane to be a semi-infinite plate of thickness 2ℓ, the multichannel reactor can be construed to be made up of a series of membranes. Thus, it can be approximated by a parallel membrane model as shown in Fig. 5.4. At steady state, the mass flux through the boundary layer is equal to that into the membrane, i.e.:

$$J = k_L [S_F - S_s] = \eta \, \ell \, k_f \, S_s \qquad [5.4]$$

where J is the steady state flux, S_f is the bulk substrate concentration, S_s is the substrate concentration at the membrane surface, and k_f is the pseudo first-order constant.

Rewriting eqn. [5.4],

$$[S_F - S_s] / J = 1 / k_L \quad \text{and} \quad S_s / J = 1 / \eta k_f \ell \qquad [5.5]$$

which leads to:

$$S_F / J = [1 / k_L] + [S_s / J] = [1 / k_L] + [1 / \eta k_f \ell] \qquad [5.6]$$

Defining a combined mass-transfer-kinetic coefficient, K' ($\equiv S_F / J$), eqn. [5.6] can be written as:

$$1 / K' = [1 / k_L] + [1 / \eta k_f \ell] \qquad [5.7]$$

Equation [5.7] represents a series of resistances, uncoupling the effects of diffusion and reaction. This is similar to the analysis of Aris (1957) for heterogeneous catalytic systems in which the reciprocal of an overall effectiveness factor is expressed as a series of resistances. In the case of spherical beads, eqn. [5.7] becomes:

$$1 / K' = [1 / k_L] + [3 / \eta k_f R] \qquad [5.8]$$

where R is the radius of the particle. For microcapsules having an aqueous phase inside a semipermeable membrane wall, the wall permeability must also be accounted for. The wall resistance $(1/k_w)$ is included as shown below (Mogensen and Vieth, 1973):

$$1 / K' = [1 / k_L] + [1 / k_w] + [3 / \eta k_f R] \qquad [5.9]$$

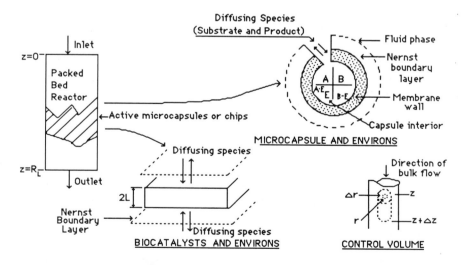

Figure 5.4   Schematic diagram of a packed-bed reactor containing spherical microcapsules or collagen-enzyme membrane in the form of chips.   z is the direction of fluid flow along the length of the reactor.   Membrane chips are considered to be semi-infinite slabs of thickness 2L. Substrate diffusion into the catalyst chips occurs in the x direction. The control volume shell is also shown.

The results obtained above - even though based on an idealized solid phase geometry - can be readily extended to immobilized enzyme reactors of practical importance, such as the spiral wound, multichannel biocatalytic modules already mentioned. Integrating the steady state material balance on the substrate passing through a differential reactor element, we obtain:

$$\ln [1 - \chi] = - K'a\tau' \qquad [5.10]$$

where a is the catalyst surface per unit of reactor fluid volume.

It is worth noting that K' can be evaluated experimentally by knowing χ and τ' for steady state reactor operation. The mass-transfer coefficient k_L is dependent on fluid velocity. Therefore, the combined coefficient K' would also be a function of fluid velocity. Experimental correlations of (K'a) with flow rate (or linear velocity) can be readily developed. An example of such a correlation is shown in Fig. 5.5 for a collagen-lactase (Eskamani, 1972) reactor system. The overall resistance (1/K'a) was found to vary linearly with the reciprocal of the flow rate. From eqn. [5.7] it is seen that the effectiveness factor η can be

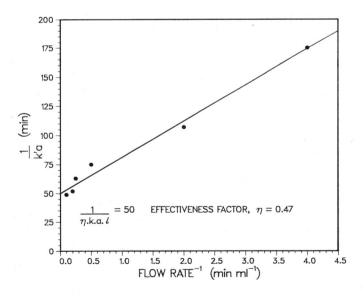

Figure 5.5   Correlation of combined mass-transfer-kinetic coefficient with flow rate, based on equation [5.7]. Experimental system studied was collagen-lactase multichannel reactor. This system obeys pseudo first-order kinetics.

evaluated from the intercept of Fig. 5.5. For instance, the effectiveness factor for the collagen-lactase system was thus evaluated to be 0.47. Correlations of this type are useful for scale-up purposes for this particular system. They can be used to design an immobilized enzyme reactor for other combinations of throughput and conversion. The reactor performance equation (eqn. [5.10]) can be written as:

$$\ln [1- \chi] = - [k_2 E_0 \tau' /K'_m] \; \eta \overline{P} \; [k_L /(k_L + [k_2 E_0 /K'_m] \eta \; \upsilon)] \qquad [5.11]$$

$$\ln [1- \chi] = - [k_2 E_0 \tau' /K'_m] \; \eta \overline{P} \; [k_L /(k_L + [k_2 E_0 /K'_m] \eta \; [R /3)] \qquad [5.12]$$

where $\overline{P}$ = La for flat sheets and (R/3)a for beads or microcapsules; it is the carrier packing factor expressed in cc of carrier per cc of fluid. Equations [5.11] and [5.12] refer to flat sheets and beads or microcapsules, respectively. The term $k_2 E_0$ is the catalytic potency of the reactor; for purposes of simplicity, let us examine the case where the dimensionless parameter $k_2 E_0 \tau'/K'_m$ is assigned the value of unity for the free enzyme in a CSTR/UF system (see column C in Table 5.1).

The quantity $k_2 E_0 \overline{P}$ can be expressed as the product of the enzyme loading factor per cc of carrier, and the reactor loading factor ($\overline{P}$) expressed in cc of carrier per cc of fluid. The values of the latter factor are shown in column D in Table 5.1. Values of the effectiveness factor η which defines the microdiffusional efficiency of the system are tabulated in column F.

In eqns. [5.11] and [5.12], the fourth term on the right-hand side represents the bulk phase transfer or "macrodiffusional" efficiency (ν) of the substrate. Column G (Table 5.1) shows this effect. Now, eqns. [5.11] and [5.12] can be rearranged in terms of the different efficiency factors as:

$$\ln (1 / [1 - \chi]) = [(1.0) \; \overline{P} \eta \nu] \qquad [5.13]$$

where (1.0) = unit value of $[k_2 E_0 \tau' / K'_m]$.

For the combined CSTR/ultrafiltration (UF) membrane reactors employing the free enzyme, a back-correction has been used; i.e., at the same level of conversion (χ = 0.9), the CSTR will require approximately two times the space-time value (τ') required for the packed bed. Therefore, relative to the latter, the CSTR has a space-time factor of 0.5, as shown in column H.

Examining columns I and J (Table 5.1), it can be noted that in the case of a free enzyme CSTR/UF process, the serious drawback of the system arises from its poor operational stability. Enzyme stability could conceivably be increased by attaching the enzyme to soluble high molecular weight supports. On the subject of stability it must also be pointed out that the operational life and characteristics of the ultrafiltration membrane itself need to be scrutinized carefully. Since the enzyme could be adsorbed on the ultrafiltration membrane, the filtration rate and the rejection efficiency could be significantly altered.

Among the immobilized enzyme reactor systems listed in Table 5.1, microcapsule reactors seem to suffer from poor stability (column I) mainly owing to breakage of capsules. More rigidity of microcapsules could possibly reduce this problem, but may increase the transport impedances. In a recent paper (Arbeloa et al., 1986), the feasibility of containing microencapsulated urease within a fluidized-bed reactor, eliminating problems of membrane rupture, was demonstrated. Hydrolysis of urea under conditions simulating that of an artificial kidney device was measured as a function of reactor residence time, microcapsule diameter, volume of microcapsules, urea concentration in the feed, and enzyme activity. Empirical correlations were developed based on dimensional analysis, which may be used to predict urea conversion within the range of experimental operating conditions. Results under transient state conditions better represent the operation of the reactor in treatment of uremic patients.

Returning to the consideration of Table 5.1, between packed columns and collagen multichannel modules, the main difference arises from the superior microdiffusional capacity of the latter, which is attributable to the very highly open internal structure of the collagen matrix - particularly under swollen conditions.

Thus, on the basis of the data appearing in column J (Table 5.1), it is clear that immobilized enzyme reactor systems can be technically advantageous as compared to free or microencapsulated systems for process scale conversion of substrates. Within the former, the biocatalytic modules employing collagen-enzyme complex membranes exhibited definite advantages of overall reactor efficiency, within the limited scope of the comparative study.

Rai (1984) has described an adaptation of the module approach for bioseparations via "cartridge chromatography." The spirally wound cartridge contains a plurality of flow compartments, providing large surface area for efficient mass interchange for protein molecules with an ion exchange matrix in membrane form. Blood proteins and enzymes have been efficiently recovered with this device.

Table 5.1 Estimate of Maximum Relative Expressed Potency of Several Single-Enzyme Reactor Configurations[a]

A Enzyme form	B Enzyme configuration	C $\dfrac{k_2 E_0 \tau'}{K'_m}$	D Carrier packing factor	E Max. relative efficiency, or C x D	F Effectiveness factor	G Macrodiffusional efficiency ν	H Contact efficiency: residence time factor relative to plug flow, 90% conversion	I Enzyme stability factor	J Maximum expressed potency relative to Case i. E x F x G x H x I
i. Free Enzyme	CSTR/UF[b]	1.0	1.0	1.0	1.0	1.0	0.5	0.01 (denaturation)	≡1.0
ii. Micro-capsules	Column	1.0	0.5	0.5	0.5	1.0	1.0	0.05 (capsule breakage)	2.5
iii. Ion Exchange Beads	Column	0.02 - 0.5	0.5	0.25	0.5	1.0	1.0	0.5 (slow leaching)	12.5
iv. Collagen Membrane	Micro-Channel Module	0.1 - 1.0	0.5	0.5	0.5	1.0	1.0	0.5 (slow denaturation)	25.0

[a] Basis: rate expression considered pseudo first-order with respect to substrate.
[b] Continuous-flow stirred-tank reactor/ultrafiltration system.

Amerace has very recently announced its modular bioreactor and separation support media. In addition to its basic flow-through modular designs based upon a microporous PVC-silica sheet, the reactor incorporates a spiral design which can be operated in a flow-over mode for processing difficult feedstocks.

5.3 MODELS FOR KINETICALLY CONTROLLED REACTIONS

In the case of intrinsically slow biochemical reactions, the mass transfer resistances sometimes become negligible. An important case in point is the isomerization of glucose to fructose by immobilized glucose isomerase (see Figure 5.6). The isomerization reaction is reversible and has an equilibrium constant of approximately 1.0 at 70°C. In our laboratory we completed a rather detailed kinetic study (Saini and Vieth, 1975). Owing to the industrial significance of this system, the analysis is briefly discussed here.

A packed bed of collagen membrane chips containing whole cells with glucose isomerase activity was used in this study. Variation of fluid mass velocity over a wide range (15 to 760 gm/cm² hr) had no appreciable effect on the initial reaction rate. Using standard mass-

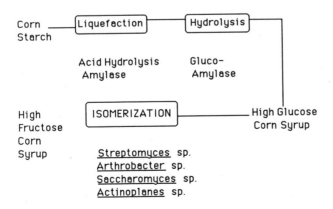

Figure 5.6   Glucose isomerization by immobilized microbes.

transfer correlations employing the Sherwood, Reynolds and Schmidt numbers, a percentage concentration drop between bulk and membrane surface was estimated to be only 0.005%. Therefore, external diffusional limitations could be ruled out. Using zero-order kinetics, the effectiveness factor was found to be 0.95, which was also corroborated experimentally. Thus, intramembrane transport resistance was also insignificant. Similar observations were made by Giniger (1973).

The isomerization reaction $A \Leftrightarrow B$ may be considered to occur in three sequential steps: (a) formation of an enzyme-substrate complex, (b) isomerization of the complex, and (c) desorption of the product. No step is considered to be exclusively rate controlling in this mechanism. The rates of the individual steps can be written as follows for the postulated reaction mechanism:

$$A + E \underset{k_{-1}}{\overset{k_1}{\Leftrightarrow}} A \cdot E \underset{k_{-2}}{\overset{k_2}{\Leftrightarrow}} B \cdot E \underset{k_{-3}}{\overset{k_3}{\Leftrightarrow}} B + E \qquad [5.14]$$

$$r_1 = k_1 [A] [E] - k_{-1} [A \cdot E] \qquad [5.15]$$

$$r_2 = k_2 [A \cdot E] - k_{-2} [B \cdot E] \qquad [5.16]$$

$$r_3 = k_3 [B \cdot E] - k_{-3} [B] [E] \qquad [5.17]$$

where k_1, k_2, k_3 are the rate constants for the forward reactions, k_{-1}, k_{-2}, k_{-3} are the rate constants for the backward reactions, $(A \cdot E)$, $(B \cdot E)$ are the reaction intermediates, and (A), (B) are the concentrations of glucose and fructose, respectively. If the enzyme E and the intermediate species $A \cdot E$ and $B \cdot E$ are assumed to be nondiffusing, the following stoichiometric invariance is obtained:

$$E_0 = E + A \cdot E + B \cdot E \qquad [5.18]$$

where E_0 is the total enzyme concentration. At steady state, all steps should be mutually rate controlling, i.e.,

$$r_1 = r_2 = r_3 = r \qquad [5.19]$$

The generalized rate expression r is obtained by solving simultaneously eqns. [5.15] - [5.19]:

$$r = \frac{m_4 E_0 [A] + m_5 E_0 [B]}{m_1 + m_2 [A] + m_3 [B]} \qquad [5.20]$$

where the lumped constants m_1 through m_5 are given by:

$$m_1 = [1/k_2] + [1/k_{-1}] + [k_{-2}/k_2 k_3] \qquad [5.21]$$

$$m_2 = [k_1/k_{-1}] ([1/k_2] + [1/k_{-3}] + [k_{-2}/k_2 k_3]) \qquad [5.22]$$

$$m_3 = [k_{-3}/k_3] ([1/k_{-2}] + [1/k_{-1}] + [k_{-2}/k_{-1}k_2]) \qquad [5.23]$$

$$m_4 = k_1/k_{-1} \qquad [5.24]$$

$$m_5 = -[k_{-2} k_{-3}/k_2 k_3] \qquad [5.25]$$

The lumped, kinetic constants were evaluated experimentally through the use of Lineweaver-Burk plots and the following rate expression is obtained on substituting these constants:

$$r = \frac{0.128 [A] - 0.098 [B]}{0.096 + 0.383 [A] + 0.25 [B]} \qquad [5.26]$$

For plug flow conditions,

$$\tau = S_0 \int_0^X \frac{dX}{r}$$

After integration, a reactor design equation is obtained as shown below:

$$[1.363 \ A_0 + 0.424] \ln [0.128 \ A_0 /(0.128 \ A_0 - 0.225 \ X)] + 0.589 \ A_0 X = \tau' \qquad [5.27]$$

Here, the fractional conversion X is defined as $(A_0 - A)/A_0$. In order to make the application of eqn. [5.27] general, certain correction factors have to be incorporated. These are necessary because (a) each batch of immobilized preparation may differ in activity; (b) contact efficiency may be different depending upon many factors, such as spacing element, length/diameter ratio of the reactor; or (c) packing density may be different.

Assuming linear dependence of reaction velocity on these three parameters, the right side of eqn. [5.27] can be modified to:

$$[\tau'] \; [\overline{P}/0.227] \; [a'/235] \; [C'/1]$$

where $\overline{P}/0.227$ is the packing density correction factor, $a'/235$ is the activity correction factor, and $C'/1.0$ is the contact efficiency correction factor. The numbers 0.227, 235 and 1.0 are, respectively, the packing density (grams of catalyst per cc of fluid volume), activity (units per gram of catalyst) and contact efficiency (observed units of activity per available (total) units of activity). With this modification eqn. [5.27] becomes,

$$[1.363 \; A_0 + 0.424] \; \ln [\; 0.128 \; A_0 \; / \; (0.128 A_0 - 0.225 \; X) \;] \; +$$

$$0.589 \; A_0 \; X \; = \; \frac{\tau' \overline{P} a' C'}{53.4} \qquad [5.28]$$

Based on the above design equation, computer simulations were performed to generate conversion-space time profiles at different feed concentrations. Experimental results agreed well with the predicted values, as shown in Fig. 5.7, for a particular combination of space time and inlet substrate concentration. More recently, the Gist-Brocades industrial process employing a lower cost variation (gelatin-immobilized whole cells with glucose isomerase activity, in bead form) has come into practice. Use of collagen enzyme systems for hydrolysis of vegetable protein (Constantinides and Adu-Amankwa, 1980) and for enhanced starch conversion efficiency (Ram and Venkatasubramanian, 1982) are other recent examples appearing in the literature.

5.4 CELL - BASED REACTORS

In approaching the analysis of immobilized whole cell reactors, appropriate consideration must be given to the two broad classifications of immobilized cell processes. Non-viable cells which express the activity of intracellular enzyme can be treated essentially as if they were carriers of a single bound enzyme system. The analysis of such a system necessitates integration of enzyme kinetics with appropriate catalyst environmental quantitative indices, (e.g., effectiveness factor) which account for external and internal mass transfer resistances as in the previous section. In contrast, for viable cells which may be resting or growing in the immobilized state, proper allowances must also be made for cell growth kinetics and cell maintenance requirements.

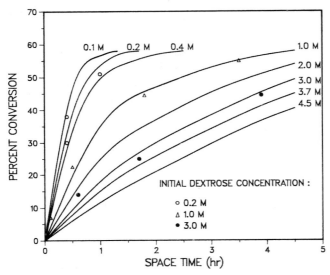

Figure 5.7 Simulation profiles and experimental data for glucose isomerization model. Numbers at the end of each curve represent the feed glucose concentration used in generating that curve.

5.5 SINGLE ENZYME TYPE IMC REACTORS

The reactor spacetime τ is defined as:

$$\tau = V_R / Q \qquad [5.29]$$

where V_R is the volume of the reactor and Q is the flow rate through the reactor.

Fractional conversion χ is expressed by:

$$\chi = [S_0 - S] / S_0 \qquad [5.30]$$

where S_0 and S are the inlet and outlet substrate concentrations respectively; reactor (volumetric) productivity P_r is defined as:

$$P_r = \chi S_0 / \tau \qquad [5.31]$$

The enzymatic rate of reaction is commonly expressed by the Briggs-Haldane monoenzyme, monosubstrate, stationary state model:

$$r = -[\,ds\,/\,dt\,] = \frac{k_2\,ES}{K_m + S} \qquad [5.32]$$

where r is the reaction rate, S is the substrate concentration, K_m is the Briggs-Haldane (popularly known as the Michaelis-Menten) constant, and $k_2\,E$ is the maximum reaction rate (V_m) for that system. Occasionally K_m and V_m are replaced by apparent constants K_m' and V_m' to account for external influences on intrinsic kinetics.

Enzyme inactivation kinetics can be adequately described as a pseudo first order process.

$$-[\,dE\,/\,dt\,] = k_d\,E \qquad [5.33]$$

where E is the effective enzyme concentration in the reactor at time t and k_d is the first order decay constant.

5.6 EFFECT OF MASS TRANSFER ON THE PERFORMANCE OF IMMOBILIZED CELL REACTORS

External (film) diffusion, diffusive and electrostatic effects, internal (pore) diffusion, and combinations of these effects constitute the array of possible influences which may be encountered.

EXTERNAL FILM DIFFUSION

Mass transfer of the substrate from the fluid to the catalyst surface can be represented by:

$$r_m = k_L\,a_m\,[S_F - S_S] \qquad [5.34]$$

where k_L is the mass transfer coefficient, a_m is the surface area for mass transfer, and S_F and S_S are the substrate concentrations in the bulk and at the surface, respectively. Correlations may be consulted to estimate k_L for different particle geometries and operating conditions (Vieth et al., 1976).

First order kinetics constitute a reasonable approximation to enzyme reaction behavior for many engineering calculations. For packed bed geometries, an equation of the type:

$$\tau' = k_f \left(- \ln [1 - \chi] \right) \qquad [5.35]$$

can be employed where $k_f = V_m'/K_m'$, the pseudo first order rate constant, and τ' is the reactor spacetime based on reactor fluid volume:

$$\tau' = [V_R \, \in] / Q \qquad [5.36]$$

The constant k_f is a limiting form of k'_f, the modified pseudo first order constant incorporating both kinetic and diffusional resistances.

$$k'_f = \frac{k_f \, k_L \, a_m}{k_f + k_L} \qquad [5.37]$$

DIFFUSIVE AND ELECTROSTATIC EFFECTS

Boundary layer diffusional resistance can appear together with substrate partitioning by electrostatic forces. Surface immobilized whole cells might be expected to display the net negative or positive charge borne on the cell walls. Perhaps it will prove possible to carry over the analysis of Hamilton et al. (1973) who examined a wide range of surface potentials using the Gouy-Chapman potential distribution.

INTERNAL (PORE) DIFFUSION

The classical approach to this problem is through the use of an effectiveness factor, η, which compactly expresses the ratio of the observed reaction rate to that which would apply if the enzyme particle were gradientless in substrate concentration. The actual situation for a single catalyst element can be described by a second order differential equation:

$$D_e \left[\partial^2 S / \partial z^2 \right] - r = 0 \qquad [5.38]$$

The solution of this equation can be put in the form of a relation for the effectiveness factor in terms of the Thiele modulus ϕ:

$$\phi = l \left[k_{true} \, S_s^{\,m-1} / D_e \right]^{0.5} \qquad [5.39]$$

where z is the distance from the center of the catalyst particle, L is the characteristic particle dimension, D_e is the effective diffusivity, m is the reaction order and k_{true} is the undisguised or true kinetic constant.

The effectiveness factor for spherical particles in a packed bed is then:

$$\eta = [1 / \phi] [1 / \tanh 3 \phi - 1 / 3 \phi] \qquad [5.40]$$

while that for a packed bed of membrane-like chips is:

$$\eta = [\tanh\phi / \phi] \qquad [5.41]$$

Once the effectiveness factor is calculated, the actual rate expression is multiplied by this factor and used in the reactor performance equation. More intricate examples, including hollow fiber ones, are described in detail in a recent publication (Venkatasubramanian et al., 1983).

5.7 IMMOBILIZED LIVING CELL SYSTEMS

Immobilization of microorganisms on the surface of solid supports has been practiced for centuries in the manufacture of vinegar. It is only within the past fifteen years that live cell immobilization has been studied for the purpose of developing more productive bioreactor systems (Hattori, 1972). Initial cell entrapment studies used synthetic polymers such as polyacrylamide; however, the immobilization conditions were extremely harsh, and natural hydrocolloid gels were soon utilized, giving much more satisfactory results (Chibata, 1979). These materials, such as agarose, agar, collagen, calcium alginate and κ-carageenan, are, with the exception of collagen, all long chain polysaccharide polymers derived from seaweed. Agarose and agar undergo gelation when the solution temperature falls below approximately 40-45°C. A solution containing sodium alginate gels in the presence of calcium ions, while carageenan gels in the presence of potassium. Immobilization using alginate is simpler than with carageenan because in the liquid state alginate hydrocolloid suspensions are much less viscous, making for easier handling.

The history of laboratory experimentation with immobilized cells (IMC) is nearly as brief as that of recombinant cell technology. The potentials which have been uncovered in the past ten to fifteen years make it an attractive technique for widespread adoption in large scale processing. There is a definite technological driving force that will lead to the implementation of IMC reactor technology, to take its place alongside classical fermentation processing. Some of the advantages of live-cell immobilization are listed below.

    i. Very high cell loadings can be obtained in the immobilized phase. In some cases, immobilized cell concentrations may be as much as one hundredfold those of free-cell concentration (Karkare et al., 1985).

    ii. IMC reactors can be operated at dilution rates well beyond the maximum specific growth rate of the microorganism without incurring cell washout.

    iii. Significantly increased productivities of biomass and product formation can be achieved.

    iv. Increased yield ratio of product formation to cell growth provides higher carbon conversion efficiencies (Karkare et al., 1986).

    v. Short and long-term reactor stability to system perturbations is improved.

    vi. In comparison to batch fermentors, IMC reactors have lower working volumes. Substantial capital savings, especially for new plants, can be realized from reduced size requirements.

    vii. IMC bioreactors can achieve a physiological uncoupling between cell growth and secondary metabolite formation by feedstock alternation; i.e., one feed for cell growth and the other for metabolite production.

5.8 ANALYSIS OF LIVE CELL REACTORS

In the following analysis (Venkatasubramanian et al., 1983), immobilized cell replication is considered to occur on the surface of the catalyst only. Oxygen limitation usually regulates maximum growth near the catalyst surface of available area, A (m^2/l), a characteristic of the supporting material used. A maximum biomass loading capacity, X_s* (g dry cell wt/m^2), can be defined; it is a function of the support as well as the microorganism used. The cell surface concentration is defined as X_S (g/m^2) and the bulk concentration is given by X (g/l). On the basis of reactor volume, cell concentration becomes $X_{im} = X_s \cdot A$ (g/l).

As usual, the reactor volume is V_R (liters) and substrate flow rate is Q (l/min). S is the concentration of limiting substrate in the reactor; S_0 at the inlet. A modification of the standard approach taken to analysis of a classical continuous culture system suffices; it was suggested in part by the work of Topiwala and Hamer (1971). A CSTR type of configuration is considered; the analysis also applies to relatively shallow fluidized beds.

Regarding the surface as the cell "hatchery," with a specific growth rate μ_S,

$$\mu_S = \frac{1}{X_S} \frac{dX_S}{dt} = \frac{\mu_{MS} S}{K_S + S} \qquad [5.42]$$

The next logical step is for surface growth to continue until catalyst loading capacity is reached. The data presented by Jirku et al. (1981) suggests that μ_S is different (lower) from the bulk growth rate, μ_b.

Once the catalyst is completely loaded, the cells *release* into the bulk with specific growth rate, μ_b, which depends on the substrate concentration:

$$\mu_b = \frac{1}{X_{total}} \frac{dX_{total}}{dt} = \mu_m \cdot \frac{S}{K_S + S} \qquad [5.43]$$

where $X_{total} = X_{im} + X$.

Live/growing cells remain attached to the catalyst while the dead cells wash away; this is supported by some recent experimental evidence which shows the preferential leaching of dead cells (Karkare et al., 1981).

5.9 IDEALIZED REACTOR PERFORMANCE EQUATIONS

For a steady state mass balance, the catalyst is assumed to be fully loaded, i.e., $X_{im} = X_S*A$. Taking inlet cell concentration to be zero, we have for a CSTR:

$$\mu_b X V_R + \mu_b X_S* A V_R = Q X \qquad [5.44]$$

Let the *dilution rate* $D = \dfrac{Q}{V_R} = \dfrac{1}{\tau}$

$$D \cdot X = \mu_b [X + X_S* A] \qquad [5.45]$$

or $\qquad D \cdot X = \mu_b \, (X + X_{im})$ $\qquad\qquad\qquad$ [5.46]

A balance on the limiting substrate provides:

$$D \, (S_0 - S) = \frac{\mu_b}{Y} \, [X + X_{im}]$$

[5.47]

where Y is the *biomass yield coefficient*.
Combine eqns. [5.46] and [5.47] to obtain:

$$X = Y \, [S_0 - S]$$

[5.48]

Equation [5.48] is identical to the analogous continuous culture equation. However, the value of S is always lower in the case of IMC reactors as shown below:
Substituting eqn. [5.48] in [5.47], and [5.45] in [5.47], one obtains:

$$\frac{\mu_m s}{K_S + S} = \frac{DY \, [S_0 - S]}{Y \, [S_0 - S] + X_{im}}$$

[5.49]

Equation [5.49] is quadratic in S and has *one* meaningful root between zero and S_0. In the limit as $X_{im} \rightarrow 0$, we regain the familiar continuous culture relationship:

$$\mu_b = D$$

[5.50]

The important result of this analysis is that $D > \mu_b$ at all values of D, implying that the exit substrate concentration is always lower for the IMC reactor at any given dilution rate. As a result, the exit cell concentration for the immobilized cell process must always be higher and, consequently, the IMC reactor is superior to the continuous culture system in terms of conversion efficiencies. Figure 5.8 shows the effect of increasing X_{im} on the exit biomass concentration. The calculations are for growth of *A. suboxydans*. The relevant constants were obtained from the data of Vera-Solis (1976). It is apparent that the washout condition in submerged continuous cultures is eliminated by using immobilized cells.

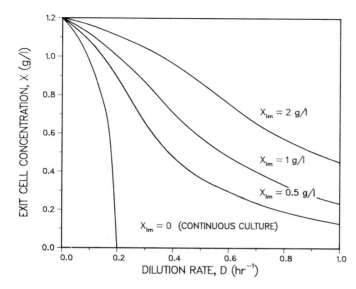

Figure 5.8 Effect of dilution rate on effluent cell concentration in immobilized live cell CSTR with varying amounts of immobilized cells. Calculations were based on data of Vera-Solis (1976) for growth of *Acetobacter suboxydans* on ethanol. μ_m = 0.23 hr-1, K_S = 6.5 g/l, Y = 0.0265 g cell/g ethanol and S_0= 45 g/l ethanol.

For the special case of $D = \mu_m$ we have a unique solution for equation [5.49].

$$S = \frac{K_S\ Y\ S_0}{K_S\ Y + X_{im}} \qquad [5.51]$$

Again, the equation reduces to $S = S_0$ as $X_{im} \to 0$. Rearranging equation [5.49] we obtain:

$$D = \frac{\mu_m\ S\ Y\ [S_0 - S] + \mu_m\ S\ X_{im}}{Y\ [K_S + S]\ [S_0 - S]} \qquad [5.52]$$

Multiplying this by $X = Y\ (S_0 - S)$,

$$DX = P_r = \frac{\mu_m\ S\ Y\ [S_0 - S] + \mu_m\ S\ X_{im}}{K_S + S} \qquad [5.53]$$

For maximum productivity w.r.t. substrate concentration:

$$dP_r / dS = 0 \quad \text{and} \quad d^2 P_r / dS^2 \Big|_{S_{opt}} < 0 \qquad [5.54]$$

Differentiation and simplification lead to:

$$Y S_{opt}^2 + 2K_S Y S_{opt} - K_S [Y S_0 + X_{im}] = 0 \qquad [5.55]$$

Using the meaningful positive root,

$$S_{opt} = \frac{\sqrt{K_S^2 Y^2 + K_S Y [YS_0 + X_{im}]} - K_S Y}{Y} \qquad [5.56]$$

and

$$D_{opt} = \frac{\mu_m S_{opt} Y [S_0 - S_{opt}] + \mu_m S_{opt} X_{im}}{Y [K_S + S_{opt}] [S_0 - S_{opt}]} \qquad [5.57]$$

It is readily verified that as $X_{im} \to 0$, equations [5.56] and 5.57] reduce to their counterparts in continuous culture systems.
For the example in Figure 5.8, considering $X_{im} = 2g/l$,

$$D_{opt} = 0.77 \text{ hr}^{-1}$$

This is much greater than μ_m; i.e., the productivity at this dilution rate is 0.463 g/l·hr, or about 3.5 times the maximum productivity of a similar continuous culture. The exit cell concentrations for the two cases are quite comparable (X = 0.6 g/l for IMC reactor and X = 0.8 g/l for continuous culture). It is obvious that the cell concentration can be increased for the IMC reactor at the expense of a little productivity and the productivity levels will still be much higher than the continuous culture. In a recent study, Okita and Kirwan (1986) have reached much the same conclusions.
Figure 5.9 shows the change in biomass productivity with dilution rates for various immobilized cell concentrations.

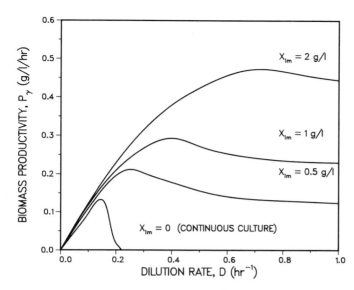

Figure 5.9 Effect of dilution rate on biomass productivity of immobilized live cell CSTR with varying amounts of immobilized cells. Calculations based on data of Vera-Solis (1976) for growth of *Acetobacter suboxydans* on ethanol. Parameter values are the same as in Figure 5.8.

5.10 CONCEPT OF A DUAL COLONY OR HYBRID REACTOR

The foregoing considerations lead us to the concept of a highly productive biomass generator (Messing et al., 1981). If a large amount of cells can be immobilized in a small volume (by using highly porous supports), biomass can be generated at much higher rates than previously possible. This is the argument for use of extended surfaces (Topiwala and Hamer, 1971) in fermentation. Since, for the production of biomass at enhanced rates we are utilizing cell growth both in surface culture and submerged culture, it is appropriate to term these reactors in particular as *hybrid reactors*. This term takes on added significance in the case of immobilized recombinant cells (described in a later chapter) where the overall gene pool is divided between the two colonies. Thus, hybrid reactors are a combination of the concepts of cell immobilization and continuous culturing of microorganisms, with a type of controlled or regulated release providing the communication between the two cell populations. In general, wherever live immobilized cells are used for continuous fermentation (whether for growth associated or non-growth associated products), the term structured bed fermentation (Vieth, 1979) seems appropriate in the design sense.

METABOLITE PRODUCTION

In general, metabolite production in fermentation systems can be described by the Leudeking-Piret model:

$$\frac{dP}{dt} = K_1 X + K_2 \frac{dX}{dt} \qquad [5.58]$$

Where P is the metabolite concentration; if $K_2 \gg K_1$ we have a growth-associated product (or primary metabolite), and when $K_1 \gg K_2$ we have a secondary metabolite.

a.   Primary   (growth-associated)   metabolites. In this case the rate of product formation is given by:

$$\frac{dP}{dt} = K_2 \frac{dX}{dt} \qquad [5.59]$$

and the metabolite productivity is described by:

$$P_r = \frac{dP}{dt} = K_2 DX \qquad [5.60]$$

Hence, maximizing the metabolite productivity is the same as maximizing biomass productivity. Therefore, the calculation of optimum dilution rate is the same as before.

b.   Secondary   metabolites. Rate of biosynthesis of secondary metabolites is dependent primarily on the cell concentration. Hence we can write this rate as:

$$\frac{dP}{dt} = K_1 X \qquad [5.61]$$

For an immobilized-cell process, this would become:

$$\frac{dP}{dt} = K_1 [X + X_{im}] \qquad [5.62]$$

Hence, maximizing the productivity in this case involves maximizing X_{im}. This can be done by using a catalyst support with high loading capacities. It is also advantageous to keep X as high as possible. From Fig. 5.8 we can see that X does not change drastically until D becomes greater than μ_m. Thus any value of D would yield substantially the same productivity. Therefore, the choice of D would depend on the yield requirement of the process. The product mass balance in this case is given by:

$$DP = K_1 [X + X_{im}]$$

or

$$P = \frac{K_1}{D} [X + X_{im}] \qquad [5.63]$$

Because X is relatively constant, D can be chosen to suit the requirement of P (often dictated by the recovery process).

c. **Metabolites with mixed growth model.** When K_1 and K_2 are both significant, the product mass balance becomes:

$$DP = K_1 [X + X_{im}] + K_2 \mu_b [X + X_{im}] = [K_1 + K_2 \mu_b] [X + X_{im}] \qquad [5.64]$$

Again, using the same techniques as in biomass productivity, we can calculate optimum dilution rate for maximum productivity. In this case,

$$S_{opt} = \frac{\sqrt{Y^2 K_S^2 [K_1 + \mu_m K_2]^2 - YK_S [K_1 + \mu_m K_2][YS_O + X_{im}] - YK_S [K_1 + \mu_m K_2]}}{Y [K_1 + \mu_m K_2]} \qquad [5.65]$$

and, once again,

$$D_{opt} = \frac{\mu_m S_{opt} Y [S_O - S_{opt}] + \mu_m S_{opt} X_{im}}{Y [K_S + S_{opt}][S_O - S_{opt}]} \qquad [5.66]$$

Again, we can verify that as $K_1 \rightarrow 0$, one obtains S_{opt} identical to that of growth-associated products.

5.11 MASS TRANSFER CONSIDERATIONS

Until recently, rather little information was available in the literature on the analysis of mass transfer effects in immobilized living cell systems. However, the treatment of bulk and pore diffusional effects up to the surface of the immobilized cell would proceed in the same manner as for the case of immobilized single enzyme type reactor systems. Next we have to consider the diffusion of the substrate and the product through the barrier imposed by the cell envelope (i.e., cell wall and cell membrane) itself.

It can safely be assumed that substrate transport into the microbial cell is characterized by passive diffusion in the case of monoenzyme-type immobilized cell systems. For such reactions the permeability of the cell envelope can sometimes be increased by specific treatments; e.g., heat treatment of cells containing glucose isomerase activity prior to immobilization (Vieth and Venkatasubramanian, 1976). With more complex reaction sequences and pathways, the total cell structure needs to be retained intact. Thus, the role of transport resistances through the cell envelope in the overall reaction assumes greater importance in the case of immobilized living cell systems.

5.12 SUBSTRATE TRANSPORT INTO THE IMMOBILIZED CELLS

The type of diffusion mechanism would depend on the organism and the limiting substrate itself. Active transport seems to be a common mode of transport of sugars into cells. Hence, models based on this concept should be used to take into account the cell envelope resistance in the reactor performance calculation. As will be shown in a later chapter, Vieth et al. (1982) have described lactose transport through the cell membrane by an equation of the type:

$$\frac{d\,[L_{in}]}{dt} = \frac{A\,(t)\,L_0}{B\,(t) + L_0} = J_L/V_C \qquad [5.67]$$

where L_{in} is the intracellular lactose concentration, L_0 is the extracellular concentration, and $A(t)$ and $B(t)$ are slightly time-dependent coefficients in batch culture. J_L is the molar flux. In the case of continuous systems, A and B are constants.

Another important transport issue relates to oxygen transfer into the cells in the immobilized state in the case of aerobic organisms. Adequate oxygen supply is ensured in traditional fermentations through the use of properly designed aeration and agitation systems. However, it is not always feasible to extrapolate this approach to immobilized cell reaction systems.   Wang (1986) and Papoutsakis (1986) have discussed the necessity of low shear oxygenation in relation to the culturing of relatively fragile mammalian cell systems. New approaches (e.g., membrane oxygenators) are now coming into play with cell reaction systems.   As alluded to already in Chapter 3, membrane oxygenator systems can be designed with predictable $k_L a$ values which can exceed those which are achievable by conventional high shear transfer techniques.

Immobilized cell bioreactor technology has already found a home in the biotechnology industry for monoclonal antibody production using hybridoma cells.   Immobilized cell bioreactors are ideally suited to this purpose for the following reasons:

1. The fragile hybridoma cells are protected in the immobilized phase from the damaging effects of shear forces present in the liquid environment.

2. The immobilized cell environment closely approximates that of natural tissue, producing high cell density while maintaining adequate levels of nutrient mass transfer.

3. IMC bioreactors can be operated for long time periods (weeks to months) while maintaining excellent cell viability and protein productivity levels.

Verax uses a gentle fluidization technique to assure adequate oxygenation in its recently described (Dean et al., 1986) licensed process for culturing mammalian cells immobilized in specially treated collagen/silica matrices.   The use of hyperbaric oxygen and/or recycle stream oxygenation furnish two other possibilities for enhancement of oxygen transfer to growing cultures which are oxygen-limited.

The foregoing discussion points to the important fact that immobilized cell systems colonize into thin layers near carrier surfaces where they operate at the verge of starvation with respect to a limiting substrate.   Chotani (1984) studied ethanol production via yeast cell culture immobilized in particles of calcium alginate in a specially baffled plug-flow like reactor (see Table 5.2).   Allowing for both substrate and product inhibition, the equation for the rate of change of substrate along the reactor length becomes:

$$-\frac{dS}{dt} = \frac{\eta \nu_{max} S}{K_{S0} + S + S^2/K_{S2}} \left[1 - P/P_m \right] \qquad [5.68]$$

where $P_m = 95$ g/l, $K_{S0} = 1$ g/l, $K_{S2} = 450$ g/l and ν_{max} is the specific growth rate limit for the yeast cells (Chotani, 1984). The diffusion-limited conditions imposed by the membrane-like carrier envelope limit fermentation activity to the outer shell of the particles, resulting in low effectiveness factors ($0.1 \leqslant \eta \leqslant 0.2$). For the sake of argument, does this imply a poor reactor design? One might just as well say that open tubular heterogeneous reactors employing thin annular layers of active biocatalyst have a low effectiveness factor on the basis of apparently "unused" tube volume. Obviously this is nonsense, as these reactors operate very efficiently in their design modality (i.e., linear analytical chemistry with rapid response). Likewise, the entrapment of yeast cells was found to exercise better control on the growth nutrients supply to the immobilized colony, which can be maintained at levels in the range of 5×10^9 cells/ml within a typical bead. As a consequence, close to theoretical glucose to ethanol conversion efficiency and large doubling time resulted for such cells. So, while the cells did not replicate efficiently, they channeled carbon to the desired product, ethanol, quite efficiently.

Table 5.2   Comparisons of Immobilized Cell Bioreactors (Chotani, 1984)

Fixed Bed	Slurry-Type Fluid-Agitated Bed	Cross-Current
No mixing	Good mixing	Intermediate mixing
Low fermentation rate	High fermentation rate	High fermentation rate
Unstable operation	Stable operation	Stable operation
Low effectiveness factor	Higher effectiveness factor	Higher effectiveness factor
Radial temperature	Good temperature control	Good temperature gradients control
Solids-free feed	Suspensions processable	Solids-free feed
Low axial dispersion	High axial dispersion	Intermediate axial dispersion

Table   5.3   Selected   Conditions   for   Analysis

Day of Operation	Dilution Rate (h^{-1})	Mode of Operation	Phosphate Concentration in Medium (moles/l)
18	0.145	steady, growth	10^{-3}
21	0.49	steady, growth	10^{-3}
27	0.149	steady, nongrowth	0
39	0.149	transient, growth	5×10^{-4}

A further, rather pointed example of this effect appears in the work of Karkare et al. (1986).   In this case, $PO_4^=$ was used as a metabolic switch in production of a secondary metabolite, candicidin, synthesized by *Streptomyces griseus* in a nongrowth mode.   Results are summarized in the data presented in Tables 5.3, 5.4 and 5.5. It was concluded that the appropriate reactor operation strategy   should   be to stimulate growth of cells to their maximum extent at regular intervals, mean- while using a nongrowth medium to enhance reactor productivity and efficiency.

5.13 MEMBRANE BIOREACTOR-SEPARATORS AND COLLAGEN TECHNOLOGY

In   its   original   approach   to   "combined-duty"   membrane appliances, our laboratory chose to focus on membrane-immobilized enzymes and cells (Vieth et al., 1973; 1974a, b; 1976a, b; 1986), arranged in a spiral wound configuration (Vieth et al., 1974a) where substrate flow passes over the thin membrane layers, as indicated in Figure 5.10.   The separation aspect was the relatively static one of the freely diffusing substrates and products vis-a-vis the fixed biocatalyst. A key scale-up parameter is the ratio of membrane surface to reactor volume.   With   proper   design,   an   effective   internal   fluid   circulation pattern can be achieved at the expense of rather modest pressure drops (Chatterjee and Belfort, 1986).   More recently, we have begun to examine multimembrane configurations which operate in a flow-through mode (Vasudevan et al., 1987), as discussed in the following section.

Table 5.4 Data for Carbon Balance

Day	Phosphate (M)	Glucose In (g/l)	Glucose Out (g/l)	Asparagine (g/l)	Cells Out (g/l)	Exit CO_2 (%)	Liquid Flow (1/h)	Gas Flow (1/min)	Candicidin Out (mg/l)
18	0.001	45.04	42.16	5.99	1.548	0.0815	0.0134	0.16	12.675
21	0.001	45.04	45.04	6.30	0.715	0.1075	0.04	0.16	6.375
27	0	43.42	42.52	7.0	0.117	0.01	0.01	0.08	24.0
39	5×10^{-4}	43.42	39.63	5.80	3.35	0.092	0.014	0.14	60.30

Table 5.5 Carbon Balance and Conversion Efficiencies for Candicidin Synthesis by Immobilized Cells

Mode	Glucose Carbon Consumed (g/h)	Asparagine Carbon Consumed (g/h)	Carbon to Candicidin (g/h)	Carbon to Cells (g/h)	Carbon to CO_2 (g/h)	Carbon Conv. Eff. (%)	Carbon Unaccounted (%)
Continuous growing cells	0.0154	0.0195	0.0027	0.0103	0.0115	7.73	32
Continuous growing cells	0.0041	0.0538	0.0041	0.0143	0.0276	7.04	22
Continuous nongrowing cells	0.0038	0.0109	0.0038	0.0006	0.0058	25.85	30.6
Continuous unsteady state growing cells	0.0211	0.0214	0.0134	0.0234	0.0057	31.52	---

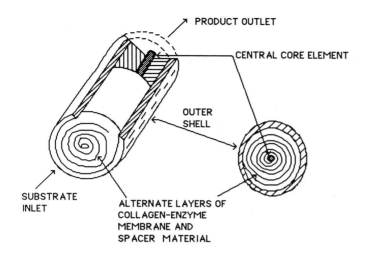

Figure 5.10   Spiral-wound microchannel biocatalytic module.

5.14 SIMULTANEOUS REACTION AND SEPARATION IN A MEMBRANE BIOREACTOR INTRODUCTION

Attempts have been made to immobilize microorganisms within membranes in order to perform bioreaction and separation simultaneously (Cho and Shuler, 1986; Tharakan and Chau, 1986; Vieth, Wang and Gilbert, 1972; Stanley and Quinn, 1987; Leeper and Tsao, 1987). The major problems encountered are high diffusional resistances, substrate depletion and product inhibition. Over the past few years, a number of membrane reactors have been proposed in which devices were constructed to solve these problems either partially or completely (Karel and Robertson, 1985; Venkat et al., 1985).

We have approached the problem with a twin objective: (1) to overcome substrate and product diffusional resistances by forcing the nutrient solution through the cell mass, and (2) to simultaneously separate product from the reaction mixture so as to reduce product inhibition. In order to accomplish the foregoing objective, a new concept of membrane reactor is presented and experimental results are analyzed in terms of the interaction between membrane diffusion and biocatalytic reaction.

Furthermore, in all previous attempts, commercially available membranes have been used to form the barrier material behind which the microorganisms are confined. As the performance of the bioreactor was thus limited by the microstructure of the membranes available in the market, the effect of membrane microporosity on the course of the biocatalytic reaction has not been fully investigated. Hence, another objective of our work is to study the effect of membrane microporosity throughly by using laboratory-made membranes, prepared under various controlled conditions.

MEMBRANE "SANDWICH" REACTOR CONCEPT

The biocatalyst is sandwiched between an ultrafiltration (UF) membrane and a reverse osmosis (RO) membrane as shown in Figure 5.11. The UF membrane separates the cells from the feed volume, while providing free passage for all nutrients to the cell mass below. The RO membrane, while immobilizing the cells, also helps in separating the nutrients (glucose and salts) from the product stream preferentially (dePinho et al., 1987). This improves product purity and concentration. An additional advantage is that the substrate uptake becomes more efficient. The cell layer thickness can be minimized in this configuration in order to prevent diffusional gradients in the cell mass due to substrate depletion. The permeate disengages at the bottom surface of the RO membrane and is collected below

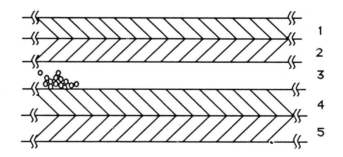

Figure 5.11 Membrane sandwich. 1. UF membrane, 2. Coarse filter paper (>10μm), 3. Cell mass, 4. Fine filter (0.2 μm, microporous), 5. RO membrane.

The materials used in the experiment and the experimental methods have been described previously (Vasudevan et al., 1987). *Saccharomyces cerevisiae* ATCC 4126 was employed as the biocatalyst because extensive immobilization studies have been done in the past on this microorganism (Chotani, 1984). Inoculum was prepared using YM broth. A measured volume (from 20-60 ml) of known cell concentration (in the exponential phase, about 2×10^7 cells per ml) was filtered through a 0.2-μm microfilter and then sandwiched between the membranes. Cellulose acetate membranes (UF and RO) were prepared in the laboratory; details are shown in Table 5.6. Pore sizes in the UF membranes were controlled by the ethanol content of the gelation media, while those of the RO membranes were controlled by the shrinkage temperature (Sourirajan and Matsuura, 1985).

Table 5.6   Details of Membrane Preparation

	UF* membranes	RO** membranes
Casting solution composition, wt%		
1. Cellulose acetate (E-398-3)	17.0	
2. Acetone	69.2	
3. Magnesium perchlorate	1.45	
4. Water	12.35	
Temperature of casting solution, °C	4	
Temperature of casting atmosphere, °C	room	
Humidity of casting atmosphere, %	room	
Solvent evaporation period, s	60	
Gelation medium	ethanol-water	water
Temperature of gelation medium, °C	0	
Shrinkage temperature, °C	- - - -	60-82
Shrinkage period, min	- - - -	10

*pore size controlled by ethanol content in gelation media.
**pore size controlled by shrinkage temperature.

A schematic diagram of the bioreactor is shown in Figure 5.12. After the membrane sandwich was mounted at the bottom of the reactor, the feed solution containing glucose and other nutrients was loaded and pressurized with nitrogen gas. The product stream was collected over time intervals ranging from 6-24 hours, based on permeaton rate, and cumulative concentrations of glucose and ethanol were measured using a Glucose Analyzer (YSI Model 27 Industrial Analyzer) and Gas Chromatograph (Hewlett-Packard 5880A series) equipped with a 1.82-meter-length glass column filled with 5% Carbowax 20M and 80/120

Carbopak B-AW.   Cell concentrations were measured using an emocytometer (grid volume-0.001 ml) by counting under a microscope. By staining the cell sample with methylene blue, cells that appeared blue were considered inactive.   Viable cells were counted using an agar plate count after incubating for 24-48 hours.

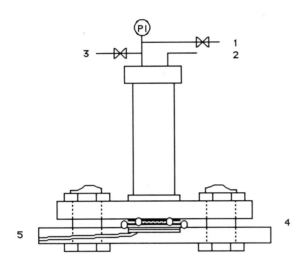

Figure 5.12 Membrane reactor. 1. Inert gas inlet, 2. Feed port, 3. Bleed port, 4. Membrane laminate, 5. Permeate collection port.

QUASI-STEADY STATE ANALYSIS

The analysis is based on the assumption that local concentrations and biocatalytic reaction rate are time-independent.   The analysis was conducted in order to know the direction of change in the membrane permeation rate when the variables involved in the biocatalytic operation (such as membrane porosities [both UF and RO], operating pressure, and feed glucose concentration) were changed.   Even though it is difficult to reach a steady state in the actual system, it was thought that this analysis would simulate the process at any given instant of time as long as the time taken for the solute concentrations to reach local stasis is much smaller than the rate of change of biocatalytic activity.   One may also infer from the experimental data that the system approaches a true steady state towards the end of the reaction period.

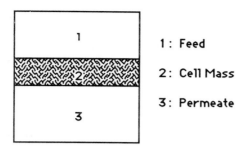

Figure 5.13 Definition of subscripts.

Assuming a steady state glucose transfer rate through the UF membrane, we can derive a set of transport equations that are analogous to those of reverse osmosis transport (Sourirajan and Matsuura, 1985).

$$N_B = A_R c \, [\mathscr{P} - B \, (cX_{GL,2} - cX_{GL,3})] \qquad [5.69]$$

$$N_B = ck_{UF,GL} \, [1 - X_{GL,3}] \, \ln \frac{[X_{GL,2} - X_{GL,3}] + \dfrac{r_g/[\mathscr{S}/V_2]}{N_B}}{[X_{GL,1} - X_{GL,3}] + \dfrac{r_g/[\mathscr{S}/V_2]}{N_B}} \qquad [5.70]$$

$$N_A = ck_{RO,GL} \, [X_{GL,2} - X_{GL,3}] - r_g/[\mathscr{S}/V_2] \qquad [5.71]$$

$$X_{GL,3} = \frac{N_A + r_g/[\mathscr{S}/V_2]}{N_A + N_B + r_g/[\mathscr{S}/V_2]} \qquad [5.72]$$

where N_A = solute flux emerging from the UF membrane (on a weight basis), g/cm^2 hr,

N_B = solvent flux through the RO membrane (on a weight basis), g/cm^2 hr,

$X_{GL,2}$ = mass fraction of glucose in the cell mass space,

$X_{GL,3}$ = mass fraction of glucose in the product permeate.

Equation [5.69] implies that the solvent flux is controlled by the RO membrane. Equation [5.70] is analogous to the equation for concentration polarization in the RO system and determines the glucose concentration in the cell mass region. Convective flow through the UF membrane is intrinsically included in the concentration polarization model. Equation [5.71] is concerned with the glucose transport through the RO membrane. Equation [5.72] is an expression for the glucose mole fraction in the permeate in terms of solute and solvent fluxes. It should be noted that the term for the reaction rate has to be included in equations [5.70] - [5.72], while in the RO transport without bioreaction, this term is "missing." Note also that the RO membrane is permeable to ethyl alcohol but the osmotic pressure of ethyl alcohol has no appreciable effect on the transport of solvent water in equation [5.69].

Therefore, one obtains four equations and four unknowns (N_A, N_B, $X_{GL,2}$ and $X_{GL,3}$). After combining equations [5.69], [5.70], [5.71] and [5.72] one can calculate N_B, provided numerical values are supplied for the pure water permeability constant for the RO membrane, A_R; the glucose transport coefficients in the UF and RO membranes, $k_{UF,GL}$ and $k_{RO,GL}$, respectively; and $r_g /(\mathcal{S} /V_2)$, the glucose consumption rate by cells, r_g, together with ($\mathcal{S} /V_2$), the surface to volume ratio. In this manner, both membrane transport and biocatalytic reaction are characterized.

5.15 EXPERIMENTAL STRATEGIES AND TRENDS

The results of typical runs are shown in Figures 5.14, 5.15 and 5.16. There is a maximum in glucose concentration in the permeate at the beginning of each run. The drop in the glucose concentration was followed by a much slower but steady decline until the end of the experiment. The ethanol concentration in the permeate always began at modest values and increased steadily, as can be seen in the figures. These observations are presumably due to cell activation being induced gradually after the introduction of a sudden change in the cell environment at the very beginning of the experiment. This change of environment is concerned with two effects - a sudden increase in cell concentration while filtering, and the need for the cells to adapt to an anaerobic means of survival. An apparent oscillatory trend in the concentration of both glucose and ethanol in the permeate has also been observed, as shown in Figure 5.16. The fact that the membranes are not pressure-treated leads to a lower initial glucose separation for the same

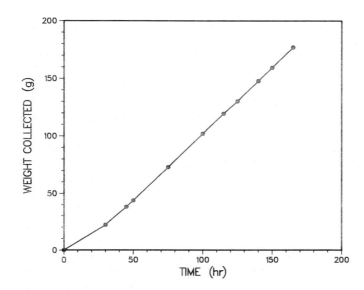

Figure 5.14   Solvent permeation.

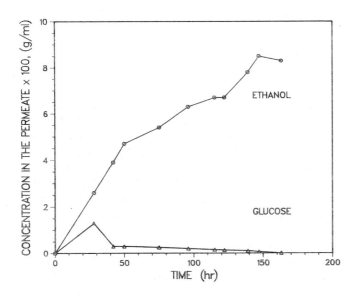

Figure 5.15   Substrate rejection and product enrichment. [UF 50%, RO 72°C, Initial cell number 1.1 x 10^9, Operating pressure 2068 kPag (=300 psig).]

product rate.   This is consistent with the observed permeation rate, which drops initially and then remains reasonably constant throughout the reaction.

Although the general pattern of the biocatalytic reaction depicted in the figures remained the same among a set of experiments, data levels change significantly with change in the operational variables. For example, as summarized in Table 5.7, the operating pressure and the feed glucose concentration have effects on the biocatalytic reaction. Though the initial total cell number in the sandwich space was maintained at a value of 1×10^9 , the cell number at the end of the run increased from 1.7×10^9 to 9.1×10^9 when the pressure was increased from 1379 to 2068 kilo Pascal guage pressure, or kPag (200 to 300 psig); this is presumably due to a better supply of nutrients at higher pressures.   The number, however, declined to 7.7×10^9 at 2758 kPag (400 psi).   The increase in the feed glucose concentration did not increase the final cell number significantly.   The maximum glucose concentration (which corresponds to the peak in glucose concentration in Figure 5.16) increased with increase in both the operating pressure and the feed glucose concentration, which seems to be natural.   The maximum ethanol concentration (obtained usually toward the end of the experiment) was in the range of 0.083 to 0.0985 g/ml and no significant correlation with the operating variables was found for it.

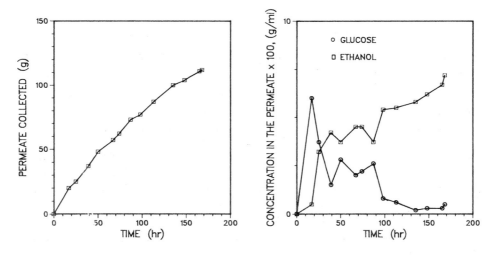

Figure 5.16   Solvent permeation (left plot), substrate rejection and product enrichment.   [UF 50%, RO 72°C, Initial cell number 1.82×10^8, Operating pressure 827 kPag (=120 psig).]

Because the ethanol concentration is in a narrow range at the end of the biocatalytic reaction whence the system has reached quasi-steady state, the ethyl alcohol production rate depends primarily on the membrane permeation characteristics of both UF and RO membranes. Therefore, an analysis of the effects of the operating variables on the permeation rate was attempted, using equations [5.69] - [5.72]. The results for the parameters are:

$A_R = 1.09 \times 10^{-4} \sim 5.0 \times 10^{-4}$ cm / [h $\cdot$ kPa] ,

$k_{UF,GL} = 0.116 \sim 0.3$ cm /h,

$k_{RO,GL} = 0.216$ cm /h,

$r_g / [\mathscr{S}/V_2] = 0.025$ g/[cm$^2 \cdot$ h].

The A_R and $k_{UF,GL}$ values were chosen so that experimental data on the amount of the permeate solution collected and the permeate glucose concentration could be approximately regenerated by numerical calculation. The $k_{RO,GL}$ value is from the literature (Matsuura and Sourirajan, 1971), and it corresponds to membranes cast under the conditions given in Table 5.6 and shrunk at 72°C. The value for ($r_g / [\mathscr{S}/V_2]$) was obtained from the permeation rate of 0.125 g/[cm$^2 \cdot$ h] of the solution that contains 10% (by weight) ethyl alcohol. It should be noted that the A_R value is much lower than the literature value (Matsuura and Sourirajan, 1971) and that $k_{UF,GL} < k_{RO,GL}$, with both results being unexpected. These values reflect a complicated inter- action between the cell mass and the membrane transport that has not yet been completely clarified.

Table 5.7   Summary of Some Experimental Results**

Operating pressure	Feed glucose conc.	Initial cell number	Final cell number	Maximum glucose conc. in permeate	Maximum ethanol conc. in permeate	Average* flux
kPag(psig)	g/ml	x 10^{-9}	x 10^{-9}	g/ml	g/ml	g/h
1379(200)	0.125	0.985	1.67	0.0001	0.0873	0.246
1724(250)	0.125	1.10	5.33	0.0085	0.0985	0.781
2068(300)	0.116	1.10	9.08	0.0143	0.086	1.104
2758(400)	0.125	1.10	7.70	0.0167	0.0975	0.699
2068(300)	0.15	1.10	5.04	0.0212	0.083	0.484
2068(300)	0.20	1.10	9.72	0.0312	0.093	0.252

*effective film area, 8 cm^2.
**Shrinkage temperature of RO membrane, 72°C;   Ethanol content in gelation medium of UF membrane, 50% (v/v).

From the data one may conclude that the optimum permeation rate can be achieved at an operating pressure of 2068 kPag (=300 psig) with a feed glucose concentration of 0.125 g/ml when an ultrafiltration membrane prepared in a gelation medium containing 35% (v/v) ethyl alcohol is used. Though the permeation rate may become higher at a lower feed glucose concentration, the ethanol concentration in the permeate, on the other hand, becomes too low to obtain a sufficiently high ethanol production rate. A lower shrinkage temperature in the RO membrane preparation would lead to higher flux but the separation of glucose would be greatly reduced. This reduction would not be desirable in order to maintain a high substrate consumption efficiency as a larger amount of glucose would tend to pass through. With this requirement as a criterion, RO membranes shrunk at 72°C were considered to be the most appropriate for achieving both high glucose rejection and high permeation rate, simultaneously.

One of the problems usually faced in the fermentation of glucose to ethanol is the removal of carbon dioxide generated as a byproduct of fermentation. While the evolved carbon dioxide inhibits cell growth, it should be noted that it does not affect ethanol production until very high partial pressures of carbon dioxide are reached (Jones and Greenfield, 1982; Norton and Krauss, 1972). This reasoning is most probably borne out in the fact that the ethanol production rate keeps increasing through the run. Once the solubility limit was reached in the cell space in all previous configurations, gaseous carbon dioxide pressure began forcing the nutrient solution out of the cell mass (Cho and Shuler, 1986; Inloes et al., 1983). In this study, it was felt that the effect of carbon dioxide was negligible. An advantage of this system is that the solvent solution is forced across the cell mass and, hence, dead spaces do not exist, which otherwise lead to carbon dioxide collection. (Likewise, in a continuous cross-flow configuration, the problem of carbon dioxide removal can be eliminated completely by stripping the nutrient recycle externally.)

One very interesting result is the finding that biocatalytic activity is not lost despite the severe environment that the cells are exposed to in terms of the higher operating pressure than those seen normally. [In a previous study (Norton and Krauss, 1972) the maximum pressure was 1379 kPa (200psig).]

5.16 PROCESS FEASIBILITY

i.The newly proposed immobilized live cell-membrane reactor has proven to be workable. Considering that a high productivity of greater than 70 g ethanol/l cell mass · h can be achieved, this bioreactor is currently at least as productive as any other immobilized cell bioreactor.

ii.Reactor productivity does not drop with time due to carbon dioxide collection because dead spaces are eliminated by forcing the nutrients across the cell mass.

iii.There are ranges of microporosity for both the UF and RO membranes where the ethanol production rate is optimized.

iv.There are ranges of operating pressure and feed glucose concentration where the ethanol producton rate is optimized.

v.More accurate numerical values for membrane transport parameters (particularly when membranes are interacting with the cell mass) are necessary.

Further studies are being conducted to better understand the process and to generate transport and kinetic parameters that will assist with optimization of the system.

5.17 NOMENCLATURE

A	Surface area available for immobilization, m^2/l
A_R	Pure water permeability constant of RO, cm/h · kPa
a_m	Surface area for mass transfer, cm^2/g catalyst
B	Proportionality constant between osmotic pressure and solute concentration, kPa · ml/gm
c	Total concentration, on weight basis, including solvent and solute (same as density of the solution), g/ml
D	Dilution rate, hr^{-1}
D_e	Effective diffusivity, cm^2/sec
D_{opt}	Optimum dilution rate for maximum volumetric productivity, hr^{-1}
E	Enzyme concentration in the reactor, g/l
E_{diff}, E_{kin}	Reactor efficiency for diffusion-controlled and kinetically controlled reactions, respectively
J	Mass flux, g/cm^2 min
J_L	Lactose molar flux into cells, moles/g cell sec
K'	Combined mass transfer-kinetic coefficient, cc of reactor fluid/cm^2 (catalyst surface) min

K_1	Non-growth associated product constant, hr^{-1}
K_2	Growth associated product constant
K_m	Briggs-Haldane/Michaelis-Menten constant, g/l
K'_m	Apparent Michaelis-Menten constant, g/l
K_S	Monod equation constant - "Half growth velocity constant", g/l
k	Membrane mass transfer coefficient, m/hr
k_2	Reaction rate constant, min^{-1}
k_d	Enzyme decay constant, min^{-1}
k_f	Pseudo first order rate constant, min^{-1}
k'_f	Modified pseudo first order rate constant, min^{-1}
k_L	Mass transfer coefficient, cm/sec
L_{in}	Intracellular lactose concentration, $moles/l$
L_o	Extracellular lactose concentration, $moles/l$
l	Characteristic dimension of catalyst particle, cm
m	Order of reaction
N_A	Solute flux through the UF membrane on weight basis, $g/cm^2\text{-}hr$
N_B	Solvent flux through the RO membrane on weight basis, $g/cm^2\text{-}hr$
$\mathcal{P}$	Operating pressure, kPa
P	Product concentration, g/l
P_r	Reactor volumetric productivity, $g/l/hr$
$\overline{P}$	Catalyst packing density, cc catalyst/cc reactor fluid
Q	Flow rate through the reactor, l/min
r	Reaction rate, $g/l/min$
r_g	Glucose consumption rate, $g/ml\text{-}hr$
r_m	Rate of mass transfer, $g/min/g$ catalyst
S	Substrate concentration, g/l
S_F	Bulk substrate concentration, g/l
S_{opt}	Optimum substrate concentration for maximum productivity, g/l
S_O	Inlet substrate concentration, g/l
S_S	Substrate concentration at the catalyst surface, g/l
$\mathcal{S}$	Effective membrane area, cm^2
t	Time, min
V	Volume, ml
V_c	Specific volume of cells, l/g cell dry weight
V_m	Maximum reaction rate, $g/l/min$
V'_m	Apparent maximum reaction rate, $g/l/min$
V_R	Reactor volume, l
X	Effluent cell concentration, g/l

X_{im}	Immobilized cell concentration expressed in terms of reactor volume, g/l
X_S	Cell concentration on the surface of support, g/m^2
X_S*	Maximum possible cell concentration on the surface of support, g/m^2
X_{total}	Total cell concentration in immobilized cell reactor including free and immobilized cells, g/l
$(cX_{GL,1})^0$	Initial glucose concentration in the feed, g/ml
χ	Fractional conversion of substrate
χ_o	Conversion at time t = 0
χ_t	Conversion at time t
$Y_{P/S}$	Product yield coefficient, g product/g substrate

GREEK LETTERS

ϵ	Fractional void volume of packed-bed reactor
ζ	Dimensionless reactor length for reactors with porous annular walls
η	Effectiveness factor
μ_b	Specific growth rate of cells in bulk, hr^{-1}
μ_m	Maximum specific growth rate of cells in bulk, hr^{-1}
μ_s	Specific growth rate of cells on the surface of the support, hr^{-1}
μ_{sm}	Maximum specific growth rate of cells on the surface of the support, hr^{-1}
ν	Macrodiffusional efficiency factor
τ	Reactor space-time = V_R/Q, min
τ'	Reactor space-time based on reactor fluid volume = $V_R \epsilon/Q$, min
ϕ	Thiele modulus = $l(k_{true} S_S{}^{m-1})/D_e)^{0.5}$
ϕ_m	Modified Thiele modulus = $l(V'_m/[K'_m D_e])^{0.5}$

SUBSCRIPTS

1,2,3	Feed solution, cell mass space and product permeate solution, respectively
ET, GL	Ethanol and glucose solute, respectively
RO	Quantities concerning RO membranes
UF	Quantities concerning UF membranes

ABBREVIATIONS

MLR	Membrane laminate reactor
RO	Reverse Osmosis
UF	Ultrafiltration

5.18 REFERENCES

Arbeloa, M., R.J. Neufeld and T.M.S. Chang, *J. Mem. Sci.*, **29**, 321 (1986).

Aris, R., *Chem. Eng. Sci.*, **6**, 262 (1957).

Bernath, F.R. and W.R. Vieth, in "Immobilized Enzymes in Food and Microbial Processes," Plenum Press: New York, p.157 (1974).

Bernath, F.R., L.S. Olanoff and W.R. Vieth, in "Biomedical Applications of Immobilized Enzymes and Proteins," Plenum Press: New York, p.351 (1976).

Bowski, L., P.M. Shah, D.Y. Ryu and W.R. Vieth, *Biotechnol. Bioeng. Symp.* No. 3, 229 (1972).

Chatterjee, S.G. and G. Belfort, *J. Mem. Sci.*, **28**, 191 (1986).

Chibata, I., in "Immobilized Microbial Cells," ACS, Symp. Series (1979).

Chibata, I., T. Tosa and T. Sato, *J. Mol. Catalysis*, **37**, 1 (1986).

Cho, T. and M.L. Shuler, *Biotechnol. Prog*, **2** (no.1), 53 (1986).

Chotani, G.K., Ph.D. Thesis in Chemical and Biochemical Engineering, Rutgers U. (1984).

Constantinides, A., W.R. Vieth and P. Fernandes, *Mol. Cell. Biochem.*, **1**, 127 (1973).

Constantinides, A. and B. Adu-Amankwa, *Biotechnol Bioeng.*, **22**, 1543 (1980).

Dean, R., S.B. Karkare and K. Venkatasubramanian, *Biochem. Eng. V Abstracts*, Engineering Foundation Conferences, New York (1986).

de Pinho, M., T.D. Nguyen, T. Matsuura and S. Sourirajan, *Ind. Eng. Chem. Proc. Des. and Dev.*, in press (1987).

Eskamani, A. Ph.D. Thesis in Chemical and Biochemical Engineering, Rutgers U. (1972).

Giniger, M., M.S. Thesis in Chemical and Biochemical Engineering, Rutgers U. (1973).

Hamilton, B.K., L.J. Stockmeyer and C.K. Colton, *J. Theor. Biol.*, **41**, 547 (1973).

Hattori, R., *J. Gen. Appl. Microbiol.*, **18**, 319 (1972).

Horvath, C. and B.A. Solomon, *Biotechnol. Bioeng.*, **14**, 885 (1972).

Horvath, C. and J.M. Engasser, *Ind. Eng. Chem. Fundam.*, **12**, 229 (1973).

Horvath, C., L.H. Shendelman and R.T. Light, *Chem. Eng. Sci.*, **28**, 375 (1973).

Inloes, D.S., D.P. Taylor, S.N. Cohen, A.S. Michaels and C.R. Robertson, *Appl. Environ. Microbiol.*, **46**, 264 (1983).

Jirku, V., J. Turkova and V. Krumphanzl, *Biotechnol. Lett.*, **3**, 509 (1981).

Jones, R.P. and P.F. Greenfield, *Enzyme Microb. Technol.*, **4**, 210 (1982).

Karel, S.F., B.L. Shari and C.R. Robertson, *Chem. Eng. Sci.*, **40**, 1321 (1985).

Karkare, S.B., G.K. Chotani and K. Venkatasubramanian, Unpublished results, Rutgers U. (1981).

Karkare, S.B., R.C. Dean and K. Venkatasubramanian, *Biotechnology*, **3**, 247 (1985).

Karkare, S.B., K. Venkatasubramanian and W.R. Vieth, *Ann. N.Y. Acad. Sci, Biochem Eng. IV*, **469**, 83 (1986).

Karkare, S.B., D.H. Burke, R.C. Dean, Jr., J. Lemontt, P. Souw and K. Venkatasubramanian, *Ann. N.Y. Acad. Sci., Biochem. Eng. IV*, **469**, 91 (1986).

Kunkee, R.E. and C.S. Ough, *Appl. Microb.*, **14**, 4 (1966).

Leeper, S.A. and G.T. Tsao, *J. Memb. Sci.*, **30**, 289 (1987).

Matsuura, T. and S. Sourirajan, *J. Appl. Polym. Sci.*, **15**, 2905 (1971).

Messing, R.A., R.A. Opperman, L.B. Simpson and M. Takeguchi, U. S. Patent Number 4,286,061 (August 25, 1981).

Mogensen, A.O. and W.R. Vieth, *Biotechnol. Bioeng.*, **15**, 467 (1973).

Norton, J.S. and R.W. Krauss, *Plant and Cell Phys.*, **13**, 139 (1972).

Okita, B. and D.J. Kirwan, *Biotechnol. Progress*, **2, No. 2**, 83 (1986).

Papoutsakis, T., Proceedings, III World Congress on Chemical Engineering, Tokyo (1986).

Pedersen, H. and C. Horvath, in "Applied Biochemistry and Bioengineering," **3**, Academic Press: New York, p.1 (1981).

Pedersen, H., L. Furler, K. Venkatasubramanian, J. Prenosil and E. Stuker, *Biotechnol. Bioeng.*, **27**, 961 (1985).

Peppas, N.A. and C.T. Reinhart, *J. Mem. Sci.*, **15**, 275 (1983).

Prenosil, J.E. and H. Pedersen, *Enzyme Microb. Technol.*, **5**, 323 (1983).

Rai, V.R., U.S. - Japan Biotechnology Conference, Honolulu, Hawaii (December 1984).

Ram, K.A. and K. Venkatasubramanian, *Biotechnol. Bioeng.*, **24**, 355 (1982).

Saini, R. and W.R. Vieth, *J. Appl. Chem. Biotechnol.*, **25**, 115 (1975).

Saini, R., W.R. Vieth and S.S. Wang, *Trans. N.Y. Acad. Sci.*, **34**, 8 (1972).

Sourirajan, S. and T. Matsuura, "Reverse Osmosis / Ultrafiltration Process Principles," Natl. Res. Council of Canada: Ottawa, pp. 466, 543, 92, (1985)

Stanley, T.J. and J.A. Quinn, *J. Memb. Sci.*, **30**, 243 (1987).

Tharakan, J.P. and P.C. Chau, *Biotechnol. Bioeng.*, **2**, 329 (1986).

Topiwala, H.H. and G. Hamer, *Biotechnol. Bioeng.*, **23**, 1683 (1971).

Vasudevan, M., M.S. Thesis in Chemical and Biochemical Engineering, Rutgers U. (1987).

Vasudevan, M., T. Matsuura, G.K. Chotani and W.R. Vieth, *Separations Sci. and Technol.*, in press (1987).

Venkatasubramanian, K., W.R. Vieth and S.S. Wang, *J. Ferment. Technol.*, **50**, 600 (1972).

Venkatasubramanian, K. and W.R. Vieth, *Biotechnol. Bioeng.*, **25**, 583 (1973).

Venkatasubramanian, K., S.B. Karkare and W.R. Vieth, in "Applied Biochemistry and Bioenginering," **4**, Academic Press: New York, p.312 (1983).

Vera-Solis, F., M.S. Thesis in Food Science and Nutrition, M.I.T. (1976).

Vieth, W.R., S.G. Gilbert, S.S. Wang and R. Saini, U.S. Patent Number 3,758,396 (1973).

Vieth, W.R., S.G. Gilbert, S.S. Wang and K. Venkatasubramanian, U.S. Patent Number 3,809,613 (1974).

Vieth, W.R., S.S. Wang and S.G. Gilbert, U.S. Patent Number 3,843,446 (1974).

Vieth, W.R., S.S. Wang and R. Saini, U.S. Patent Number 3,972,776 (1976).

Vieth, W.R., S.S. Wang and S.G. Gilbert, U.S. Patent Number 3,977,941 (1976).

Vieth, W.R., S.S. Wang and S.G. Gilbert, U.S. Patent Number 4,601,981 (1986).

Vieth, W.R., S.S. Wang and S.G. Gilbert, *Biotechnol. Bioeng. Symp. Ser.*, **3**, 285 (1972).

Vieth, W.R., K. Kaushik and K. Venkatasubramanian, *Biotechnol. Bioeng.*, **24**, 1455 (1982).

Vieth, W.R., in "Biochemical Engineering," Ann. N.Y. Adad. Sci., **326**, pp.1-6 (1979).

Vieth, W.R. and K. Venkatasubramanian, in *Methods of Enzymology*, Academic Press, New York, **44**, 263 (1976).

Vieth, W.R., K. Venkatasubramanian, A. Constantinides and B. Davidson, in Applied Biochemistry and Bioengineering," **1**, Academic Press: New York, p. 221 (1976).

Wang, S.S. and W.R. Vieth, *Biotechnol. Bioeng.*, **15**, 93 (1973).

Wang, D., Proceedings, III World Congress on Chemical Engineering, Tokyo (1986).

Round on the ends and high in the
   middle,
Tell me if you know.
Don't you think that's a neat little
   riddle,
Round on the ends and high in the
   middle?
You can find it on the map,
If you look high and low.
The Os around it, HI in the middle,
   OHIO.

REACTION-TRANSPORT COUPLING IN ANAEROBIC DIGESTION: SELF-IMMOBILIZED CULTURES

6.0 INTRODUCTION

A subtle form of transport regulation of reaction pathways appears to be practiced by certain mixed cultures of anaerobic bacteria (Bhatia, 1983). Because relatively little attention has previously been paid to the subject of diffusion-reaction coupling in anaerobic systems, and because the requisite rate equations for consumption of mixed substrates and parameter estimation techniques provide a good preview in preparation for the next chapter, we go into this subject in some detail. Insight into this phenomenon is gained through process modeling and verification, as described below.

Until recently, proposed models of substrate consumption in anaerobic systems contained either simple Monod types of substrate consumption equations or slightly more complex equations, viz., substrate inhibition. Common to these approaches were two basic assumptions:

i) The substrate volatile fatty acid (VFA) consumption process is growth-associated, and

ii) The different substrates present in the feed (acetic, propionic and butyric acids) can be lumped together and reported as a composite substrate by measures such as COD (chemical oxygen demand) or BOD (biological oxygen demand).

A thorough experimental study demonstrated that feature (i) above is not valid (Bhatia, 1983). This stems mainly from the fact that in all the transient studies performed, the production rate of methane stabilized at a new steady state in less than 0.5 hours when the influent conditions were changed. Such a rapid response of the culture to the environment would be possible only if the rate controlling step in the mechanism of methane production were very close to the cell's external environment. That part of the cell which maintains its internal environment and accounts for its maintenance energy requirement, i.e., the "ATP pumps," is known to be situated in the cell membrane.

Furthermore, the individual acids present in the feed interact; i.e., inhibit each other's consumption, so feature (ii) also becomes invalid. Thus, models based on COD or BOD would not be able to account for these interactions without resorting to non-mechanistic model equations.

The above makes it obvious that a kinetic model incorporating equations for the consumption of each substrate is required. Such a model should not have to implicitly or explicitly assume that VFA consumption is a growth-associated process.

6.1 CARBON FLOW INSIDE THE CELL

The cell machinery is envisaged to function in the manner suggested by Ramakrishna (1966), Pirt (1975) and Van der Meer (1975). Carbon in the substrate is utilized to produce *energy* (primarily in the form of Adenosine Triphosphate or ATP) or is used to build the infrastructure of the cell for reproduction (see Figure 6.1). Since methane is known to be a byproduct of the *energy*-consuming metabolic pathway, the higher the energy consumption, the higher is the production of methane.

The energy produced is utilized for biochemical transformations that keep the cell's internal environment viable and for biochemical transformations necessary for the growth and replication of the cell. It is important to note the fine distinctions between various kinds of energy flow and how they lead to an unconventional definition of yield coefficient:

$$E_T = E_G + E_M \qquad\qquad [6.1]$$

where:

E_T = Total energy production (units could be number of moles of ATP),
E_G = Energy required in cellular reproduction mechanism,
E_M = Energy required by the cell to maintain its viability.

From Figure 6.1 it is shown that methane production is proportional to E_T and anything that increases it will cause methane production rate to increase. Since most of the carbon input into the cell goes into gas production, most of the carbon would be accounted for by E_T. Whatever carbon is left unaccounted for can be construed to go into

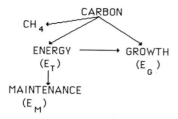

Figure 6.1 Intracellular carbon flow.

waste products or new cell mass (this is usually 5 - 10%). Thus, the biomass yield coefficient can be defined as follows:

$$Y = \frac{[\text{Total amount of carbon consumed}] - [\text{Amount of carbon consumed for gas production}]}{\text{Amount of carbon in the new cell mass}}$$

A similar definition of Y was developed by Van der Meer (1975), but since he accounted for "waste" products with a separate term, Y in his work became equal to unity.

6.2 MAINTENANCE ENERGY

Any abrupt change in the cellular environment would change the growth rate and/or the energy required to maintain the cell's viability (Ramakrishna, 1966). Thus, it is not surprising that past modeling efforts have assumed methane to be a growth-associated product. It was reasoned that, since an increase in substrate concentration increased growth rate which resulted in increased energy consumption (hence increased gas production), methane was growth-associated. Since anaerobic cultures grow very slowly (doubling time is of the order of a few days) it is difficult to conceive that they respond to substrate changes as rapidly as methane does. On the other hand, if methane is assumed to be a nongrowth-associated product then this enhanced gas production could be attributed to an increase in the maintenance energy coefficient (m); i.e., an increase in the maintenance energy requirement (E_M) of the culture. This is entirely possible since m is a very strong function of ionic strength (Pirt, 1975; Stouthamer, 1973);   i.e., as the ionic strength of the external environment increases, the cell machinery has to expend greater amounts of cellular energy to maintain its internal environment.   In anaerobic systems this greater expenditure of energy will result in greater production of methane.

In laboratory experiments where fatty acid concentration increases, an equivalent amount of base has to be added to maintain the pH.   This increases the ionic strength of the medium, causing the chain of events which leads to enhanced methane production.   Thus the following expression, appropriately modified to take into account the existence of any inhibition, can be used for maintenance energy (Ramakrishna, 1966):

$$m = \frac{M \cdot S}{K_m + S} \qquad [6.2]$$

The above expression has tacitly been used by Van der Meer (1975) and Bastin and Wandrey (1981) in their analysis of the kinetics of VFA consumption.

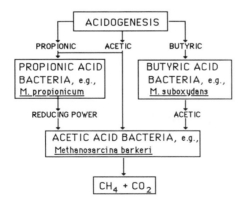

Figure 6.2 Extracellular carbon flow.

Whether methane is growth or nongrowth-associated, the model proposed in the following sections is unaffected, because most of the substrate carbon goes into gas production (90-95%), making the substrate consumption rate directly proportional to gas production rate and not to a weighted sum of growth rate and product generation rate. The model presented is, therefore, generally applicable.

6.3 SCHEMATIC DESCRIPTION OF THE MODEL

The model being proposed is designed to simulate the action of "methane-formers" in as much detail as possible. Thus, as a starting point, the block diagram proposed by Cohen et al. (1979; 1980) has been adopted.

Since most higher fatty acids are almost insoluble in water, most soured soluble organic wastes would contain lower fatty acids ($C_1 - C_3$) (Van der Meer, 1975; Mahr, 1969). Hence, it is assumed that the feed input to the reactor consists of acetic, propionic and butyric acids only. Figure 6.2 shows the flow of carbon from the three acids into and out of the different bacterial cultures. Since the oxidation of VFA mixtures requires a common enzyme for activation, it is assumed that propionic acid consuming cultures can also consume acetic acid.

Strict anaerobic (methanogenic) cultures are incapable of consuming molecules larger than C_2 (Stadtman, 1967) but they readily consume acetic acid and reduce CO_2. They are also known to exist in a symbiotic relationship with cultures that preferentially consume hydrogen (Ghose, 1975). Thus the presence of an intermediate (I in Figure 6.3) is postulated. The overall scheme of consumption is represented in Figure 6.3. The lines represent all the possible places that each acid could exert its inhibitory influence. All the possible interactions, mapped out in Figure 6.3, do not necessarily occur but they have been included for completeness. Additionally, the mathematics developed to describe the system (next section) will use this general form of interaction while the following section employs experimental data to drop the statistically insignificant interactions.

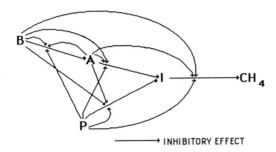

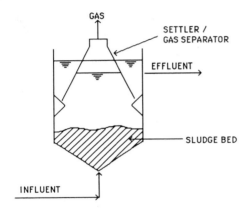

Figure 6.3   Model schema for substrate flow.

Figure 6.4   Upflow anaerobic sludge bed reactor.

6.4 REACTOR SET UP

Since the upflow anaerobic sludge bed (UASB) system responded like a CSTR to flow rate step-ups, the entire reactor has been modeled as a CSTR (Bhatia, 1983). Van der Meer in his doctoral dissertation (1975) found that the UASB system responds to tracer step and/or pulse inputs as two CSTRs in series. One CSTR represents the biomass bed (Figure 6.4) and the other CSTR represents the disengaging space. This model of the reactor system was used in writing the system equations in the next subsection.

6.5 MODEL EQUATIONS
SUBSTRATE CONSUMPTION TERM

Using the notation developed by Roels and Kossens (1978) one can write the following general equations necessary for conducting a mass balance on component i of the liquid stream.

$$r_i = \frac{R_i \cdot S_i}{[1 + K_{Si} \cdot S_i] \cdot [1 + \sum_{j=1}^{3} K_{Iij} \cdot S_i]} \qquad [6.3]$$

where:

r_i = rate of consumption of the i-th substrate,

R_i = maximum rate of consumption of the i-th substrate,

K_{Si} = substrate affinity coefficient of the i-th substrate (similar to the reciprocal of the K_S in a Michaelis-Menten equation),

K_{Iij} = coefficient of inhibition depicting the inhibitory effect of the j-th substrate on the consumption of the i-th substrate.

It should be noted that the amount of active cell mass does not appear explicitly in equation [6.3] but does so implicitly in the term R_i; i.e., this term is a composite of the maximum specific consumption rate and the active biomass fraction consuming the i-th substrate. Since the experimental strategy was to perform the study of UASB in a manner designed to avoid growth effects, the fraction of biomass consuming a given substrate can be taken to be a constant.

Additionally, in case substrate inhibition, as suspected, exists, one can depict this situation using equation [6.3] wherein the inhibition coefficients for i = j are nonzero. Thus, equation [6.3] is a very general representation of various particular substrate consumption situations (including interaction of various substrates).

MECHANISMS CAUSING MULTIPLE STEADY STATES

The manifestation of multiple steady states is an indication that at low substrate concentrations the rate controlling step is different from that observed at higher concentrations. The two possible mechanisms that can explain such a phenomenon are discussed in the following two sections.

SUBSTRATE INHIBITION

In the case of simple substrate inhibition of an enzyme or a rate controlling sequence of reactions inside a cell, the following is the mechanistic representation:

$$E + S \Leftrightarrow ES \rightarrow E + P \qquad [6.4]$$

$$ES + S \Leftrightarrow ES2 \qquad [6.5]$$

where: E = enzyme or cell,
      S = substrate,
      ES, ES2 = enzyme substrate intermediates,
      P = product.

In the above case, step [6.4] is rate controlling at low substrate concentrations and the product generation rate exhibits a Monod or Michaelis-Menten type of functionality. As the substrate concentration increases, the reaction represented in equation [6.5] moves more and more towards the right of the equilibrium. This increasingly inhibits the ES-complex and causes the production rate, which is proportional to the concentration of the complex, to decrease. Thus, at very high substrate concentrations, a switchover of rate controlling mechanisms occurs; i.e., the product generation rate now becomes a monotonically decreasing function of the substrate concentration. This reaction scheme can lead to the manifestation of multiple steady states (Bhatia, 1983).

6.6 HYSTERESIS DUE TO STRUCTURE OF THE BEADS

The UASB system is not set up like a conventional chemostat; i.e., the cultures present therein are not well mixed but rather are flocculated. The granular nature of the catalyst can lead to the manifestation of multiple steady states due to the way the various bacterial colonies are structured inside the beads.

Butyric acid-consuming bacteria are *aerobic* and/or facultative. Thus, in the UASB system, these bacteria are predominantly found on the outer layers of the biomass particles. Similarly, since the propionic acid bacteria are *anaerobic*, while methanogenic bacteria are strictly *anaerobic*, the model envisages such bacteria to form layers as shown in Figure 6.5. In view of the fact that the biomass studied had been adapted to an undeaerated feed, it is highly plausible that over the protracted process of adaptation the aerobic bacteria clustered on the outside of the particle while the anaerobes accumulated inside.

Shinmoyo et al. (1982) have shown that carageenan-immobilized aerobic bacteria growing inside the matrix could only penetrate 50 microns of the catalyst particle diameter. Their study was conducted in an aerated reactor in which oxygen transfer would have been less a problem than in the UASB reactor system. Additionally, Wada et al. (1980) found that immobilized yeast grown under anaerobic conditions could penetrate the beads up to 400 microns. These studies indicate that the less severe the oxygen requirement of the culture, the easier it is to penetrate deeper into the bead. Thus, in the case of a strict anaerobe, the entire population would have to reside in the core of the particle (Figure 6.5).

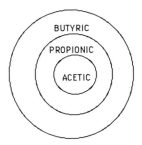

Figure 6.5 Schematic diagram of the proposed arrangement of the cultures inside the biomass particle.

The above tricolonic layering of cultures inside the particle, combined with the stoichiometry of Figure 6.3, can explain the manifestations of multiple steady states as follows:

At low concentration all the propionic acid substrate entering the bead would be readily taken up by the propionic acid-consuming bacteria. Furthermore, propionic acid is a known inhibitor of anaerobic cultures (i.e., the cultures that oxidize the reducing power generated [Figure 6.3]). Thus the layered structure of the beads would effectively protect the innermost layer of strict anaerobes from propionic acid's inhibitory effect. Beyond the threshold concentration, propionic acid cannot be consumed completely by the outer layers of bacteria and the strict anaerobes which convert the intermediate to methane are inhibited by the excess propionic acid. At this point the conversion of intermediate (I in Figure 6.3) to methane becomes the rate controlling step; i.e., switchover occurs.

6.7 DESCRIPTION OF THE REACTOR DYNAMICS

Applying the reactor balance equations to the specific case of three acid consumption in a CSTR and utilizing the stoichiometry envisaged in Figure 6.3, we have for the CSTR -I (Figure 6.6):

$$dS_1 / dt = D_L \cdot [S_{10} - S_1] - r_1 + (120 / 88) \cdot r_3 \qquad [6.6]$$

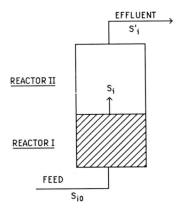

Figure 6.6 Schematic diagram of the reactor.

$$\frac{dS_2}{dt} = D_L \cdot [S_{20} - S_2] - r_2 \qquad\qquad [6.7]$$

$$\frac{dS_3}{dt} = D_L \cdot [S_{30} - S_3] - r_3 \qquad\qquad [6.8]$$

where:

S_{i0} = the concentration of the i-th substrate in the feed,

D_L = dilution rate in reactor-I CSTR; i.e., the reactor that contains the biomass,

i = 1, acetic acid,

i = 2, propionic acid,

i = 3, butyric acid.

In equation [6.6] the last term on the right hand side is added because of the stoichiometry; i.e., two moles of acetic acid (molecular weight = 60) are produced per mole of butyric acid (molecular weight = 88).

Writing a mass balance for the reactor- II CSTR we have:

$$\frac{dS'_i}{dt} = D_U \cdot [S_i - S'_i] \qquad\qquad [6.9]$$

where:

D_U = Dilution rate in the reactor-II,

S_i = Inlet concentration of the i-th substrate into the reactor-II (same as the exit concentration from reactor-I [Figure 6.6]),

S'_i = Exit concentration from the UASB system.

Since there is a negligible amount, if any, of biomass present in reactor-II, it can be safely assumed that there is no reaction occurring in the upper part. Thus the upper part has been construed as a system adding a first order lag to the transient response of the effluent from the biomass bed in the UASB system, which has no effect on the steady state response.

LAG TIMES

There is only one more facet of the model that needs to be introduced; i.e., delayed response of the "reaction machinery" or lag time. This is necessitated by the observed peaking response of the

UASB system to changes in the influent conditions. There are two kinds of lag times; (i) structural or spatial and (ii) physiological.

i)   Structural lag time is manifested due to the time it takes the change in the bulk of the reactor to register in the microenvironment of the cell population. This type of lag takes into account the mass transfer resistance. In case the biomass is not particulate, the structural lag time would be a manifestation of bulk mass transfer resistance only. In general, the lag is, of course, the analog of the penetrant time lag, already discussed.

ii)   Physiological lag time is the time taken by the cell to respond to an external change and reach its new steady state. This represents a "gearing-up and down" phenomenon for the enzymatic machinery of the cells. The manifestation of this lag would be a change in reaction rates from an initial state to the final state. Since not much is known about the nature of these responses, a first order lag response has been assumed. This is the simplest of the cases that can be considered and will be introduced in the model as follows:

$$\frac{dr_i}{dt} = D_i^* \cdot [r'_i - r_i] \qquad [6.10]$$

where:

r_i = the current rate at which the i-th substrate is being consumed,

r'_i = the calculated rate of consumption corresponding to the current substrate concentration in the bulk. Thus r'_i can be construed to be the potential reaction rate,

D_i^*= the dilution rate in the conceptualized CSTR in which the reaction is occurring.

6.8 MODEL DISCRIMINATION

Since statistical analysis of the data can fit numerous models "adequately," it becomes necessary to rationally discriminate among the various possibilities; this section is devoted to just this sort of a discrimination.

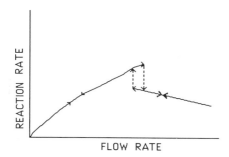

Figure 6.7   Reaction rate versus flow rate from theoretical data.

It is quite obvious that at least one of the acids exhibits a substrate inhibition behavior or some other complex mechanism. Figure 6.7 shows the plot of steady state reaction rate versus flow rate for the substrate inhibition case and Figure 6.8 is a representation of observed reaction rate versus flow rate. It is clear that even though both systems exhibit the effect of hysteresis, their responses to different flow rates are not the same. In the theoretical case the reaction rate on the right limb of the graph decreases with the flow rate while in the experimental case it rises.   This indicates that if substrate inhibition is the mechanism, then, at upper flow rates, some process is enhancing the reaction rate for propionic acid; i.e., undoing its inhibitory effect.

Thus, to explain the experimental data, it would be necessary to postulate that the right side of the curve in Figure 6.7 rises as flow rate increases. This is equivalent to saying that the ratio of the two flow rates, between which multiple steady states are possible, decreases.   In such a case, the span of the multiple steady state region would be smaller than a standard theoretical case. This is important to note since the discussion to follow will prove that the standard theoretical case does not have a large enough span to explain the data.

S_{max} is the value of substrate where the maximum reaction rate occurs. This is given by:

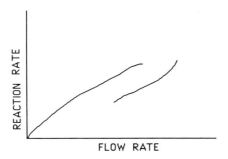

Figure 6.8   Reaction rate versus flow rate from actual data.

$$S_{max} = \sqrt{G_i \cdot H_i} = \frac{S_{iO}}{n} \tag{6.11}$$

as shown by Bhatia (1983).

The necessary condition for the existence of multiple steady states is:

$$S_{iO} \geqslant G_i + H_i + 3 \cdot [G_i \cdot H_i \cdot S_{iO}]^{1/3} \tag{6.12}$$

where G_i and H_i are the reciprocals of K_{Si} and K_{Iii}. From eqns. [6.11 and 6.12] we have:

$$S_{iO} \geqslant G_i + H_i + 3 \cdot [S_{iO}^3 / n^2]^{1/3}$$

or
$$S_{iO} \cdot [1 - (3 / n^{2/3})] > G_i + H_i \tag{6.13}$$

Since G_i and H_i are positive, the term on the left hand side of the above equation has to be positive. Thus the necessary condition becomes:

$$1 > \frac{3}{n^{2/3}}$$

or $\qquad n > \sqrt{27}$ [6.14]

In this work, for the steady state experiments, the inlet concentration was maintained at 600 ppm. Thus, according to the above analysis, S_{max} should satisfy the following condition:

$$S_{max} \leqslant \frac{S_{i0}}{\sqrt{27}}$$ [6.15]

From the above equation and the actual experimental data it becomes obvious that the maximum concentration the upper curve in Figure 6.8 could possibly reach is 115 ppm. This is not the case, as the upper curve was found to reach concentrations of up to 250 ppm. Thus a model incorporating substrate inhibition alone cannot explain the observed existence of multiple steady states.

At this point it is interesting to examine Figure 6.9, a representation of the plot of the rate of consumption of acetic or propionic acid versus the concentration of the acid. It can be seen that with increasing substrate concentration, the consumption rates of propionic acid (and acetic acid) increase. At $S_{critical}$ there is a sudden drop in rate for either acid (250 ppm for propionic acid and 120 ppm for acetic acid). It is also interesting to note that the percentage drop in reaction rates in the cases of both acetic and propionic acids is the same (about 50%). This suggests that beyond a propionic acid concentration of 250 ppm the mechanism of consumption is the same for both acids.

Further, since propionic acid has been widely reported as an inhibitor and acetic acid a reactor operation stabilizer, it is conceivable that acetic acid reduces the inhibitory effects of propionic acid. This would also explain the increasing reaction rate with propionic acid concentration after the intermediate's conversion to methane becomes the rate controlling step because the ratio of acetic to propionic acid concentration increases.

The conversion of VFA to methane is (like any other biochemical transformation) a multistep process involving various enzymes. Generally, one of the enzymes involved is appreciably slower than the others; this results in that particular enzymatic step becoming the rate controlling step. The overall kinetics of the biochemical transformation then become a manifestation of this key enzyme's kinetic character. In such a situation the other enzymatic reaction steps can be construed to be at equilibrium. From a mechanistic point of view the following sequence of reactions is proposed:

$$E + I \xrightarrow{k_1} CH_4 \qquad\qquad [6.16]$$

$$E + A \underset{k_{-2}}{\overset{k_2}{\Leftrightarrow}} EA \qquad\qquad [6.17]$$

$$E + P \underset{k_{-3}}{\overset{k_3}{\Leftrightarrow}} EP \qquad\qquad [6.18]$$

$$EA + I \xrightarrow{k_4} CH_4 \qquad\qquad [6.19]$$

where: $[E]$ = concentration of the key enzyme,
$\quad\quad\;\; [I]$ = concentration of the intermediate,
$\quad\quad\;\; [A]$ = concentration of acetic acid,
$\quad\quad\;\; [P]$ = concentration of propionic acid.

This is only a representation of the actual reactions taking place, assuming that the other reaction steps involved are at equilibrium. All the ks shown in the above sequence of reactions are composite; i.e., they incorporate ks of all the individual steps at equilibrium, as mentioned above. Equations [6.16] and [6.19] are represented as irreversible reactions because the standard free energy of formation of methane is highly negative (- 12.1 Kcal/g-mole). In addition, the gas produced is essentially insoluble in water. Thus, the reaction depicting the production of methane is essentially driven to the product side; i.e., methane.

From the above sequence of reactions it is obvious that the rate of conversion of the intermediate would be proportional to the sum of the concentrations of $[E]$ and $[EA]$; i.e., assuming that propionic acid inhibits the conversion completely.

$$\text{rate} \propto \{ [E] + [EA] \} \qquad\qquad [6.20]$$

$$[E_O] = [E] + [EA] + [EP] \qquad\qquad [6.21]$$

where $[E_O]$ = total amount of reactive sites available, and if reaction equations [6.17] and [6.18] are assumed to be at equilibrium, then:

$$\text{rate} = k_1 \cdot [E] + k_4 \cdot [EA] \qquad [6.22]$$

From eqn. [6.12]:

$$[EA] = K_2 \cdot [E] \cdot [A] \qquad [6.23]$$

where $K_2 = k_2 / k_{-2}$, and from equation [6.18]:

$$[EP] = K_3 \cdot [E] \cdot [P] \qquad [6.24]$$

where $K_3 = k_3 / k_{-3}$.
Thus, by substituting equations [6.23] and [6.24] into [6.21], one obtains:

$$[E] = \frac{[E_O]}{1 + K_2 \cdot [A] + K_3 \cdot [P]} \qquad [6.25]$$

Substituting equations [6.23] and [6.25] into equation [6.22] we have:

$$\text{rate} = \frac{[k_1 + k_4 \cdot K_2 \cdot (A)] \cdot E_O}{1 + K_2 \cdot [A] + K_3 \cdot [P]} \qquad [6.26]$$

Generally, the amount of free or unattached [E] available is negligible; i.e.,

$$[E_O] \approx [EA] + [EP]$$

so equation [6.26] becomes:

$$\text{rate} = \frac{[k_4 \cdot K_2 \cdot E_O] \cdot [A]}{K_2 \cdot [A] + K_3 \cdot [P]}$$

$$= \frac{k_4 \cdot E_O}{1 + \dfrac{K_3}{K_2} \cdot \dfrac{[P]}{[A]}} \qquad [6.27]$$

The above equation system was used to model the right limb of Fig. 6.9.

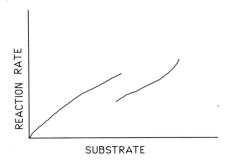

Figure 6.9   Reaction rate versus substrate from actual data.

6.9 PARAMETER ESTIMATION

The general model and the model discrimination method discussed previously are further developed in this section to apply to the specific experiments which had been performed.   Since the steady state experiments span a larger range of concentrations than transient response studies, it was deemed appropriate to estimate the parameters from the former data set.   The parameters estimated would then have a wider range of applicability.   This also simplifies the nonlinear estimation routines because steady state estimation does not involve differential equations.   Parameter estimation from steady state data was conducted by partitioning the data into two sections (see Figure 6.9). This was done for acetic and propionic acids only, since butyric acid steady state data do not indicate the existence of multiple steady states. The partitioning was done to account for the two different rate controlling mechanisms.

Now, it is clear that acetic acid inhibits the consumption of propionic acid and vice versa, while butyric acid is not inhibited by either acetic or propionic acid.   Thus in equation [6.3]:

$$K_{I12} \geqslant 0 \tag{6.28}$$

$$K_{I21} \geqslant 0 \tag{6.29}$$

$$K_{I31} = K_{I32} = 0 \tag{6.30}$$

Using the above restrictions and equations [6.3] and [6.6] - [6.10] at steady state, a computer subroutine in FORTRAN IV was developed for parameter estimation. The subroutine utilized the NONLIN computer library program, developed by UpJohn (Metzler et al., 1974), for parameter estimation.

Table 6.1   Parameter Values

Acetic Acid	Propionic Acid	Butyric Acid
$R_1 = 48.5$ (hr^{-1})	$R_2 = 12.6$ (hr^{-1})	$R_3 = 124.4$ (hr^{-1})
$K_{I12} = 4.1$ (1/g)	$K_{I21} = 19.0$ (1/g)	$K_{S3} = 64.0$ (1/g)
$K_1 = 3.93$ (g/l/hr)	$K_1 = 1.02$ (g/l/hr)	
$D_1{}^* = 0.025$ (hr^{-1})	$D_2{}^* = 0.025$ (hr^{-1})	$D_3{}^* = 0.1$ (hr^{-1})
	$S_{X\text{-over}} = 256$ (ppm)	

All parameters that did not improve the statistical fit of the data were dropped, subject to the above conditions (equations [6.28] - [6.30]).

The estimation of parameters to predict the other section of the partitioned data was performed using equations [6.27] and [6.6] - [6.10]. Table 6.1 provides the list of estimated parameters and Figures 6.10 to 6.12 display the plots of calculated effluent substrate concentration versus flow rate.

The agreement between the calculated values and experimental data is excellent, with the coefficient of correlation ranging between 0.9 and 0.98 for different substrates. Thus the reaction rate equations become:

$$r_1 = \frac{R_1 \cdot S_1}{1 + K_{I12} \cdot S_2} \qquad \text{for } S_2 \leqslant S_{X\text{-over}} \qquad [6.31a]$$

$$r_1 = \frac{K_1 \cdot S_1}{S_2} \qquad \text{for } S_2 > S_{X\text{-over}} \qquad [6.31b]$$

$$r_2 = \frac{R_2 \cdot S_2}{1 + K_{I21} \cdot S_1} \qquad \text{for } S_2 \leqslant S_{X\text{-over}} \qquad [6.32a]$$

$$r_2 = \frac{K_2 \cdot S_1}{S_2} \qquad \text{for } S_2 > S_{X\text{-over}} \qquad [6.32b]$$

$$r_3 = \frac{R_3 \cdot S_3}{1 + K_{S33} \cdot S_3} \qquad [6.33]$$

Since the above equations for reaction rates were developed using steady state data, they cannot be expected alone to predict the transient response of the system.   This necessitated the use of some of the transient data so that the D_i* in equation [6.10] could be evaluated for the system.

The D_i*, as discussed earlier, represent the composite of two types of lag processes in the biological system.   Since it is expected that the butyric acid-consuming bacteria occur in the outermost layers of the particles, D_3* (reciprocal of lag time for butyric acid bacteria) should be larger than D_1* and D_2*.   Also, D_1* and D_2* can be expected to be of the same magnitude since both cultures are anaerobic and would tend to exist in one another's vicinity.

Data presented in Figures 6.13 and 6.14 (i.e., data for butyric acid step-up and step-down)   were used to estimate the D_i*.   Equations

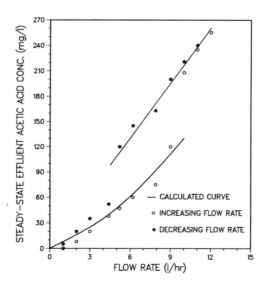

Figure 6.10   Steady state effluent values.

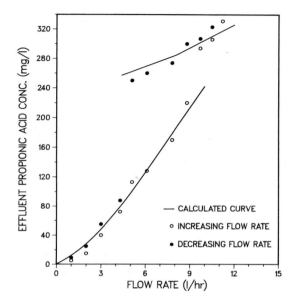

Figure 6.11   Steady state effluent values.

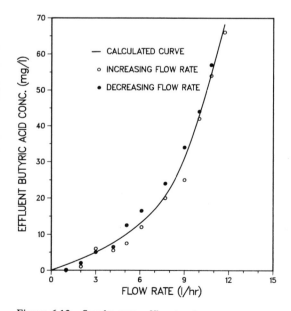

Figure 6.12   Steady state effluent values.

[6.6] - [6.10] and [6.31] - [6.33] were employed to define the system. The NONLIN program, mentioned earlier, is also capable of estimating parameters when the model is defined by a set of differential equations. It was found that the parameter estimation technique required considerable Central Processing Unit (CPU) time and the residual sum of squares was decreasing very slowly.   This necessitated the use of another library program (by UpJohn) called MAP.   This program finds the residual sum of squares at different grid points in the parameter space.   Using this program the various D_i* were estimated, thus completing the parameter estimation of the model.   The various values of D_i* are given in Table 6.1.

Figures 6.15 to 6.17 show superimposed plots of actual and predicted response of the effluent concentrations when butyric acid is stepped up and down.   The predicted response of the UASB fit the actual response for all three acids very well.   It should, however, be noted that the predicted curves do not match the data very well when the *step down* occurs;   i.e., the predicted curve dips much more than the actual data. This could be due to the fact that D_i* in the case for step-up and step-down are different.   This is in agreement with chemostat data reported in the literature; i.e., the response of a biological reactor (CSTR) peaks when inlet concentrations are stepped up but the effluent concentration decreases monotonically when influent concentration is stepped down.

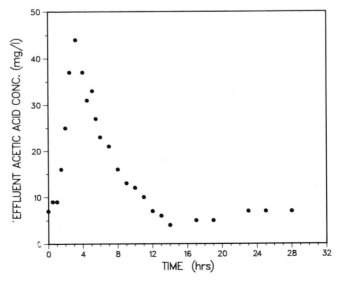

Figure 6.13   Response to 50% butyric acid step-up/down.

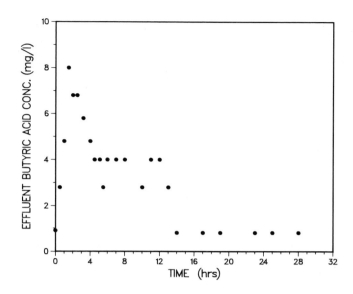

Figure 6.14   Response to 50% butyric acid step-up/down.

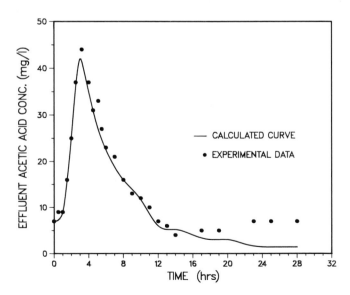

Figure 6.15   Response to 50% butyric acid step-up/down.

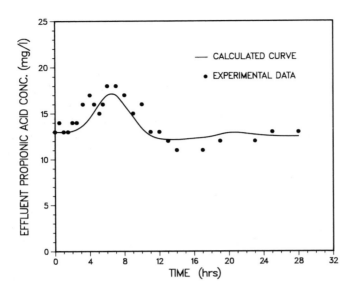

Figure 6.16   Response to 50% butyric acid step-up/down.

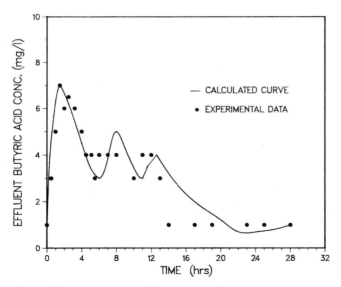

Figure 6.17   Response to 50% butyric acid step-up/down.

6.10 MODEL PREDICTIONS

Utilizing the model described in the previous sections the predicted response curves of the UASB system are generated for acetic and propionic acids step-up and step-down.   Figures 6.18 to 6.21 show the aforementioned curves with actual experimental data superimposed upon them.   Again, as in the case of butyric acid data, the predicted response very closely tracks the actual response while the step-up is in progress but when the step-down occurs the predicted response dips more than the actual data.   This again suggests the possible existence of some mechanism by which the response of the biological system is faster when the system is responding to a step-down.   In view of the fact that it takes some time for the culture to induce additional amounts of various enzymes when the substrate concentration is increased, this lag is not surprising.   In stepping down, no additional enzymes are required and the system will respond faster.

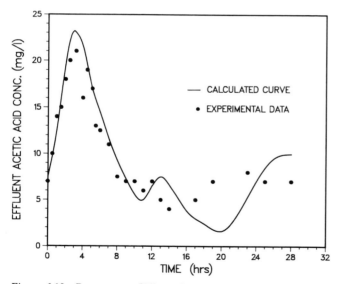

Figure 6.18   Response to 50% acetic acid step-up/down.

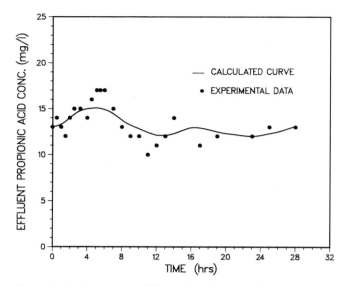

Figure 6.19   Response to 50% acetic acid step-up/down.

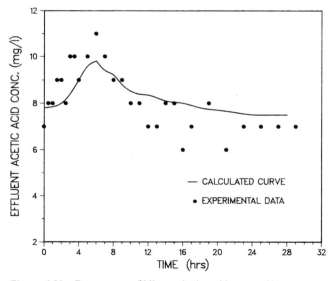

Figure 6.20   Response to 50% propionic acid step-up/down.

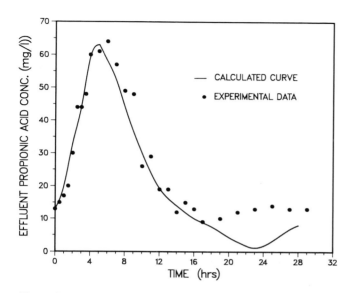

Figure 6.21   Response to 50% propionic acid step-up/down.

6.11 RECAPITULATION

In the above discussion it has been shown that the UASB system data cannot be explained by resorting to a simple Monod-type substrate consumption pattern.   A substrate inhibition model was also ruled out because the substrate concentration range over which hysteresis was observed was much larger than any substrate inhibition model could predict.

Since in the UASB system the biocatalyst is flocculated, it was found that a model incorporating this facet of the reactor setup could explain the steady state data very well.   Using the parameters generated from steady state data and data from butyric acid step change; i.e., the entire set of parameters (Table 6.1), a very good agreement between predicted and observed data was found.   International Mathematical and Statistical Libraries (IMSL) and UpJohn's NONLIN Library, combined with various root-finding and integrating subroutines, were used for parameter estimation.

The model thus described was used to predict the response of the UASB system when acetic acid and propionic acid influent concentrations were stepped up/down.   The agreement between the predicted and observed data was found to be excellent in each case during the step up schedule. During the step down the data seemed to indicate that the UASB system, like any other chemostat, responded faster than predicted. This could be due to the fact that when the culture has to "gear up," part of the lag time is the time required for the cell to produce the requisite amount of enzymes. In the case of "gearing down" this time is not required and the system responds faster.

6.12 ANALYTICAL HIERARCHIES FOR CONSIDERATION OF MICROSTRUCTURAL EFFECTS

Most bioreactor modeling efforts have characterized biocatalysts, whether free or immobilized, as uniform unsegregated colonies of cells with spatially constant properties. In the case of disgregated free cell systems operating in a chemostat, such a characterization results in relatively simple system equations and very stable transient responses to step inputs (i.e., the eigenvalues of the Jacobian matrix of the linearized system equations are real).

On the other hand, if a single microbial specie is immobilized (either on a carrier or in a network of auto-excreted polysaccharides), then the well-mixed chemostatic conditions cease to apply and diffusional aspects have to be considered as an important part of the system equations. Explicit incorporation of a diffusion equation for a limiting substrate could lead to more complicated system equations. The eigenvalues generated by such a system might be complex, leading to the possibility of oscillatory response and manifestation of multiple steady states. (The standard technique for locating such bifurcations is to look for model parameters which cause the eigenvalues (of the Jacobian matrix of the linearized equations) with largest real part to make a transition from negative to positive real root.)

The system equations become more nonlinear still when multiple-specie diffusional effects are being considered in structurally anisotropic systems; e.g., anaerobic sludge reactors. It has been shown (Bhatia et al., 1985) that analysis of the system equations in such a situation makes it necessary to take into account kinetic, biological and, at least implicitly, diffusional interactions. A ternary VFA substrate, combined with a tricolonic layering of biocatalysts, led to a cubic equation with, of course, multiple roots.

Explicitly, detailed modeling would lead to a set of nonlinear differential equations which invariably have complex eigenvalues. Nonetheless, it is this sort of analysis that must ultimately be used to analyze complicated systems like anaerobic digestors. The analysis in turn would help in elucidating the nature of spatial segregation of the various biological populations, resulting in manifestation of diffusional time delays, which invariably lead to erractic reactor response unless the reactor system is closely controlled.

6.13 REFERENCES

Bastin, K. and C. Wandrey, *Ann. N.Y. Acad. Sci., Biochemical Engineering II*, **369**, 135 (1981).

Bhatia, D., Ph.D. Thesis in Chemical and Biochemical Engineering, Rutgers U. (1983).

Bhatia, D., W.R. Vieth and K. Venkatasubramanian, *Biotechnol. Bioeng.*, **27**, 1199 (1985).

Cohen, A., A.M. Breure, J.G. van Andel and A. van Deursen, *Water Res.*, **14**, 1439 (1980).

Cohen, A., R.J. Zoetmeyer, A. van Deursen and J.G. van Andel, *Water Res.*, **13**, 571 (1979).

Ghose, S., J.R. Conrad and D.L. Klass, *J.W.P.C.F.*, **47**, 30 (1975).

Mahr, I., *Water Res.*, **3**, 507 (1969).

Metzler, C.M., G.L. Elfring and A.J. McEwen, "NONLIN and Associated Programs," The UpJohn Co., Kalamazoo, Michigan (1974).

Pirt, S.J., *Proc. R. Soc., London, Ser. B*, **163**, 224 (1975).

Ramakrishna, D., A.G. Fredrickson and H.M. Tsuchiya, *J. Gen. Appl. Microbiol.*, **12**, 4 (1966).

Roels, J.A. and N.W.F. Kossen, *Prog. Ind. Microbiol.*, **14**, 95 (1978).

Shinmoyo, A., H. Kimura and H. Okada, *Eur. J. Appl. Microbial Biotechnol.*, **19**, 7 (1982).

Stadtman, T.C., *Ann. Rev. Microbiol.*, **21**, 121 (1967).

Stouthamer, A.H. and C. Bettenhaussen, *Biochimica et Biophysica Acta*, **301**, 53 (1973).

Van der Meer, R.R., Ph.D. Thesis, Delft Univ. Press, The Netherlands (1975).

Wada, M., J. Kato and I. Chibata, *Eur. J. Appl. Microbial Biotechnol.*, **10**, 275 (1980).

You may talk of Clara Nolan's ball or anything
        you choose,
But it couldn't hold a snuffbox to the spree at
        Killigrews.
If you want your eyeballs straightened out, just come
        next week with me,
You'll have to wear your glasses at the Killigrew
        soiree!

Folk Ballad, "Killigrew's Soiree"

7

ROLE OF INTRACELLULAR CHEMICAL MESSENGER
TRANSPORT IN ENZYME BIOSYNTHESIS

7.0 INTRODUCTION

Sherwood et al. (1975) observe that, "In living tissue, diffusion may take place in the direction of a negative concentration gradient. The phenomenon of *active transport* is presumed to be due to an input of free energy or work needed for concentration by diffusion, causing a solute to diffuse *uphill*. If this process were understood it might find application in industry."

With respect to applications in biotechnology, the foregoing has a prophetic ring, as we shall see. But, to get the discussion started, a few simple examples of facilitated transport, the step just below on the transport ladder, seem to be in order. In some instances, such as acetylcholine (ACh) binding to receptor sites in the post-synaptic membrane, there exists a very large equilibrium membrane accumulation. Now, the enzyme reaction catalyzed by acetylcholinesterase is functioning to remove the penetrant "downhill" at the downstream portal of exit. In this manner, a form of facilitated transport could be sustained. Likewise, in a time lag experiment, with a glassy polymer where membrane accumulation can reach a substantial level and dual mode transport is observed, a large volume of sweep fluid removes the transported penetrant, performing a similar function.

In contrast, in the case of biosynthesis of an inducible enzyme, such as β-galactosidase, the inducer, lactose, is *actively* transported "uphill" by an enzyme-like *permease* system which functions in *concert* with a pH gradient, according to the Mitchell hypothesis (1961). Some consequences of this phenomenon are illustrated in the results shown below in Tables 7.1 and 7.2.

Table 7.1   Active Transport of Lactose in Batch Fermentation (Kaushik, 1981).

t Batch time (hr)	L_{out} Bulk Substrate Concentration, g/l	Calculated Substrate	L_{in} Intracellular Concentration, g/l
0	12.5		0
1	11.8		1.3
2	11.4		2.6
3	11.1		3.8
4	10.4		5.0
5	9.7		6.1
6	9.0		7.2
7	8.3		8.2
8	7.6		9.1
9	6.9		10.0
10	6.2		10.8

These results reveal something of the genius of evolution; i.e., the "learned" ability of the microorganism to "boot-strap" the internal concentration of an inducer-substrate (lactose) to levels where biosynthesis rates are high, even in the presence of a faltering feed supply level.

Table 7.2 Active Transport of Lactose in Continuous Culture (Ray, 1985).

D (hr^{-1}) Dilution Rate	L_{out} Bulk Lactose Concentration, g/l	L_{in} Measured Intracellular Lactose Concentration, g/l
0.1	12	10
0.2	12	39
0.3	12	45
0.4	12	61
0.5	12	61
0.6	12	48

Once again, internal levels are efficient for microbial biosynthesis except at the very lowest feed supply rates.

7.1 BIOSYNTHESIS OF GLUCOSE ISOMERASE

In some earlier mixed substrate studies (X̲ylose, G̲lucose) with the industrially-important enzyme, glucose isomerase, several peculiarities regarding its biosynthesis profiles were noted in our laboratory (Gondo et al., 1978). In the case of the single substrate-inducer, xylose, the predicted intracellular enzyme concentration is very low (Fig. 7.1, curve 6), only slightly higher than the curve for the glucose-only case (curve 7), in spite of the absence of catabolite repression in the former. However, the intracellular enzyme specific activity actually obtained experimentally for *Streptomyces venezuelae* for the xylose-only case, is higher than the maximum enzyme specific activity for the two-substrate case of equal inlet concentration (curve 3), and it is almost constant, independent of the dilution rate. The discrepancy between the experimental data and the simulation for the case of xylose as the sole carbon source can be resolved by introducing

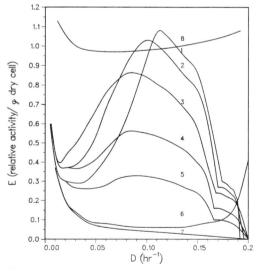

Figure 7.1   Effect of the feed bulk compositions of two substrates on the profile of the intracellular enzyme concentration.   Curve 1: $G_{Bin}=0.0025$ g/ml and $X_{Bin}=0.0175$ g/ml. Curve 2: $G_{Bin}=0.005$ and $X_{Bin}=0.015$.   Curve 3: $G_{Bin}=0.01$ and $X_{Bin}=0.01$.   Curve 4: $G_{Bin}=0.015$ and $X_{Bin}=0.005$. Curve 5: $G_{Bin}=0.0175$ and $X_{Bin}=0.0025$. Curve 6: $G_{Bin}=0.0$ and $X_{Bin}=0.02$. Curve 7: $G_{Bin}=0.02$ and $X_{Bin}=0.0$ Curve 8: $G_{Bin}=0.0$ and $X_{Bin}=0.02$.

the distribution of substrate between the outside and inside of the cell into the modeling equations. By introducing the relation $X_C = 100X_B$, the profile of the intracellular enzyme was raised, as shown in Figure 7.1, to curve No.8.

Comparisons of the theoretical curves of intracellular enzyme specific activity with experimental data for glucose isomerase biosynthesis in *Streptomyces venezuelae* are shown in Figure 7.2. Although the fit of the theoretical curve for the xylose-only case is relatively poor at low dilution rate, it can be stated that the modeling equations for catabolite repression of inducible enzyme biosynthesis derived in this work otherwise characterize the phenomena reasonably well.

7.2 INDUCIBLE ENZYME BIOSYNTHESIS: LACTOSE INDUCTION OF β-GALACTOSIDASE IN E. COLI

To solve the apparent puzzle outlined above, it was decided to carry out systematic studies with a better characterized system, and the synthesis of β-galactosidase via the *lac operon* was chosen for further work (Kaushik, 1981). Let us retrace our steps a bit, to begin at the beginning.

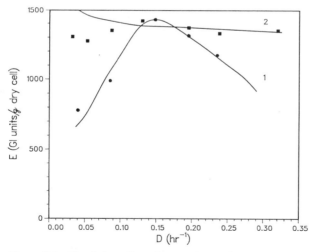

Figure 7.2 Simulation of continuous glucose isomerase biosynthesis in *St. venezuelae*. A unit of glucose isomerase (GI unit) is defined at the amount of enzyme which produces 1 mg fructose/hr at 70°C and pH 7.5. Curve 1: G_B=0.05g/ml and X_C=X_B. Curve 2: G_B=0.05 and X_C=150X_B. Feed substrate concentrations: (●) 0.7% glucose and 0.8% xylose; (■) 1.2% xylose. Solid lines are simulation curves.

Enzyme synthesis in the microbial cell is controlled by complex regulatory mechanisms. The amount of enzyme produced by a microorganism can vary dramatically, depending on the intracellular and extracellular conditions. In view of the increasing importance of enzymes as industrial catalysts, research toward understanding the biosynthetic machinery of enzyme production is clearly of major significance. In order to provide a firm basis for design, scale-up and operation of fermentations for enzyme production, there is a need for more complete models, which quantify our understanding of the regulatory mechanisms involved in enzyme biosynthesis, and which can be subjected to experimental verification.

Due to the extreme complexity of biological systems, the mechanistic knowledge available, even for well-studied systems, is usually restricted to individual steps in the overall mechanism. Relatively few attempts have been made to combine the information available on the discrete elements of the mechanisms for enzyme biosynthesis, and to formulate a comprehensive description of the overall process. By judiciously combining available mechanistic knowledge with empirical observations, reliable engineering models can be developed for each individual step in the regulatory mechanism. Ideally, every equation should represent a phenomenologically accurate description of the elementary process using established mechanistic theory, reliance on empirical correlations being kept to the minimum. Elementary models emphasizing the detailed mechanism can then be reconstituted into a set of equations that quantitatively describes the overall process and provides understanding and predictive capability. Finally, the adequacy of the mathematical model can be verified by comparing actual experimental data with model predictions.

The intent of our study was to develop a mathematical framework to model the detailed regulatory mechanisms involved in microbial enzyme biosynthesis. Specific emphasis was given to three areas in which quantitative understanding is currently limited: inducer transport, regulation of intracellular cyclic AMP (cAMP) by glucose and control of catabolite repression by cAMP levels in the cell.

In this section, the focus of discussion is on transport of inducer into the cell; models for regulation of cAMP, induction and catabolite repression have been reported in other publications (Kaushik et al., 1979; Vieth et al., 1982; Gondo et al., 1978). In the verification of the transport model presented below, the experimental conditions approximate typical fermentation conditions. Hence, induction and repression, as well as transport, can all be expected to influence enzyme production. In order to account for these effects and model the overall process, we have applied the other sub-models for these modes of regulation in conjunction with the transport model. Whereas most of the model equations used here are applicable to inducible enzymes in general, in view of the detailed mechanistic understanding recently developed on β-galactoside transport in *Escherichia coli*, the focus of the modeling efforts was the induction of β-galactosidase in *E. coli*. The relevant features of the mechanisms which regulate enzyme biosynthesis are reviewed, very briefly, below to provide a perspective for transport modeling.

7.3 REGULATION OF ENZYME BIOSYNTHESIS

β-galactosidase is synthesized in *E. coli* through an interacting sequence of mechanistic steps. A slightly oversimplified schematic diagram of the mechanisms that are invloved in this control is shown in Fig. 7.3. This includes (i) transport of glucose, lactose or other substrates across the cell membrane, and (ii) induction/repression mechanisms at the gene level, including catabolite repression.

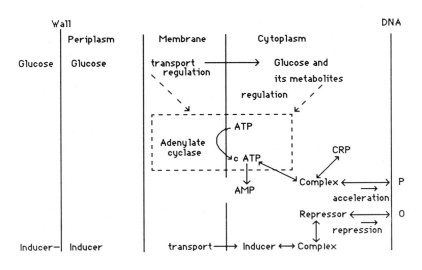

Figure 7.3 Control mechanisms involved in regulation of inducible enzyme biosynthesis. The box (---) represents the regulation of cAMP by glucose.

ENZYME INDUCTION AND REPRESSION (CYTOPLASMIC REPRESSION)

According to the model of Jacob and Monod (Monod, 1951; Jacob and Monod, 1961), the genes that determine the ability of a microorganism to make and regulate the synthesis of inducible enzyme(s) are clustered in a small region of the chromosome. The structure of the protein(s) to be synthesized is determined by the structural genes. The transcription of these genes is initiated at a single locale called the promoter gene. A group of genes thus transcribed into a single messenger RNA species is termed an operon. In the absence of an inducer of the operon, transcription of the genes is prevented by the presence of a repressor, which is coded for independently by a regulator gene. The repressor stops transcription by binding to the operator gene, which lies between the promoter and the structural gene. This is termed repression, which acts as a negative control mechanism.

In the case of induction, an inducer - which is thought to act as an allosteric effecter by changing the conformation of the repressor molecule in such a way that its affinity for the operator site is decreased - combines with the repressor, causing its release from the operator and thus allowing the transcription of mRNA.

CONTROL OF TRANSCRIPTION BY CAMP

The binding of repressor on the operator site has been discussed above. The other control site on the DNA strand lies on the promoter locus. Normally, RNA polymerase binds to its recognition site on the promoter and forms an inert complex which prevents transcription. In order to permit transcription, the RNA polymerase must be shifted from its entry site to the tight binding hepta block on the promoter (Pastan and Adhya, 1976). The RNA polymerase complex on the hepta block is called the 'open' form. It is only in this form that transcription can proceed. It is found that an equilibrium exists between the open and inert complexes.

Cyclic AMP (cAMP) exercises its control on enzyme induction by regulating the shift to the 'open' form. cAMP binds to a receptor protein (CRP) and this cAMP-CRP complex then binds to DNA, causing a conformational change, which shifts the 'inert' form of RNA polymerase-DNA to the 'open' form.

CONTROL OF INTRACELLULAR CAMP BY GLUCOSE

It is found that severe repression of enzyme induction is caused by the addition of glucose to the growth medium. This 'catabolite repression' may be correlated with the effect of glucose on the level of cAMP in the intracellular environment (Peterkofsky and Gazdar, 1974; Perlman and Pastan, 1968). cAMP is synthesized from adenosine triphosphate (ATP) by the enzyme-catalyzed reaction:

$$\text{ATP} \xrightarrow{\quad \text{adenylate cyclase} \quad} \text{cAMP} + \text{PP}_i \qquad [7.1]$$

where PP_i is pyrophosphate. The enzyme (adenylate cyclase) is bound to the plasma membrane of the cell. Abou-Sabe´ et al. (1975a, b; 1973) concluded from a variety of experiments on glucose transport that glucose interacts during transport with the adenylate cyclase system, forming an allosteric effector, thereby inhibiting its activity, and hence depressing the level of cAMP in the cell. Thus, glucose indirectly lowers the rate of enzyme production.

7.4 MODELING THE TRANSPORT OF β-GALACTOSIDES ACROSS THE PLASMA MEMBRANE

Active transport of β-galactosides (including lactose) is due to the action of a specific transport system localized in the plasma membrane (the β-galactoside permease or transport protein). This transport system has been identified by Kolber and Stein (1966; 1967) and characterized by Jones and Kennedy (1969) as the M protein. (M.W. 30,000 Daltons), with binding affinities for β-galactosides. The transport system shows high specificity toward the transported species. The presence of β-galactoside permease is essential for the growth of *E. coli* on lactose; lac mutants lacking the permease system are cryptic toward lactose (i.e., are unable to grow on lactose) (Kepes, 1964,1971). Kinetically, the transport system appears to involve the dual steps of weakly contributing passive diffusion with superimposed and overriding active transport, which is suppressed by externally added inhibitors of metabolism such as dinitrophenol (Kotyk and Janacek, 1970). The transport exhibits saturation kinetics, i.e., the substrate is accumulated up to a certain limiting concentration (up to 400 times the concentration in the external medium). Permease-positive cells may be induced by very low inducer concentrations in the bulk because the action of the permease may accumulate these low-bulk concentrations to high levels of intracellular induced concentrations.   The control of intracellular inducer concentration by the permease is therefore an important control mechanism regulating enzyme biosynthesis.

The most recent view of lacose translocation involves energy transduction in the form of proton co-transport.   Although all the details of the mechanism are not yet absolutely clear, a substantial amount of experimental effort has resulted in the elucidation of the essential features.

Lactose transport is coupled primarily to oxidation of D-lactate to pyruvate, catalyzed by a flavin adenine dinucleotide (FAD), linked with membrane-bound D-lactate dehydrogenase (D-LDH) (Kaback, 1976).   Kohn and Kaback (1973) indicate that the sequence of reactions between the primary dehydrogenase and the first cytochrome in the respiratory chain of $E.$ $coli$ (cytochrome b_1) is the region where respiratory energy is converted to work in the form of solute translocation against a chemical gradient.   In this case (aerobic respiration), electrons from D-lactate are passed to oxygen via the membrane-bound respiratory chain.   Stroobant and Kaback (1975) have shown that nonphysiological electron donors (e.g., reduced phenazine methosulfate) which donate electrons to the respiratory chain at a site prior to the cytochromes can also drive transport.   In the case of anaerobic electron-transfer systems (also shown to be membrane-bound), lactose transport can be coupled to the oxidation of α-glycerol-P with fumarate as an acceptor or the oxidation of formate with nitrate as an electron acceptor.

To date, the exact mechanism by which oxidation is coupled to solute transport has not been established.   However, there is a great deal of experimental evidence to indicate that transport is coupled to the (electrochemical) components of the proton gradient (Kaczorowski and Kaback, 1979; Kaczorowski et al., 1979).   Mitchell's chemiosmotic hypothesis (1961, 1966, 1972, 1973) postulates the generation of a proton gradient caused by the expulsion of protons to the medium when oxidation of electron donors occurs via the membrane-bound respiratory chain.   A similar proton gradient could also be developed by ATP hydrolysis catalysed by a membrane-bound ATPase.   In either case, once the proton gradient is generated, it is used as the generalized energy source for solute translocation, the specificity of each solute transport system being governed by its affinity characteristics to use a particular component of the electrochemical gradient.   The electro-chemical potential gradient is composed of electrical and chemical components given by:

$$\Delta \overline{\mu}_{H+} = n \; [\Delta \psi - 2.3 \; [RT/F] \; \Delta pH] \qquad [7.2]$$

where $\Delta \psi$ is the electrical potential drop across the membrane and ΔpH is the difference in the logarithms of the proton concentrations across the membrane.   In short, lactose is thought to be transported through cotransport with protons and coupled to $\Delta \overline{\mu}_{H+}$. The extent of lactose accumulation (-195 mV) can be thermodynamically reconciled to the electrochemical potential gradient (Kaback, 1976) (ΔpH = - 120mV, $\Delta \psi$ = - 75mV at pH 5.5, and n = 1).

The kinetics of lactose translocation are determined by the rates of the binding and translocation steps involved in the mechanism for transport.   If the rates of the elementary steps are disproportionately different, the kinetics can be approximated by a rate-determining step. Figure 7.4   shows schematically a generalized sequence of kinetic steps that seems to correspond most closely to available experimental evidence on lactose-proton symport.   The 'carrier,' represented by C, carries a net negative charge in the free state.   The sequence is general in that the reactions are viewed as transitions between states.   Physical movement of the carrier is not implied.   The reaction sequences 1 - 3

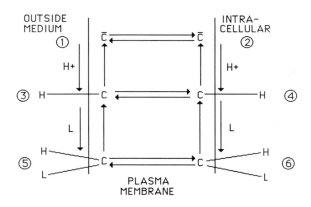

Figure 7.4   Schematic kinetic sequence for lactose translocation. $\overline{C}$ represents   the carrier site; C-H, C-L-H represent carrier bound to a proton (H^+) and to a proton and a lactose molecule (L), respectively.

and 2 - 4 represent the binding of H^+ outside and inside the membrane, respectively. Similarly, 3 - 5 and 4 - 6 represent lactose binding. For the efflux of lactose, the net reaction sequence is 4 - 6 - 5 - 3 - 1 - 2 and for influx it is 3 - 5 - 6 - 4 - 2 - 1. The reaction step 3 - 4 dissipates the proton gradient without lactose transport and its retarding effect is probably quite small under normal conditions.

If N is the total number of sites available for binding and translocation of the lac carrier, we can denote the number of sites at any time t in each state (1 through 6) as N_1, N_2 ... N_6. Assuming, further, that the binding reactions are first order with respect to site concentrations and solute concentrations (the subscript 'o' denoting outside and the subscript 'in' denoting inside),

$$\alpha_{13} = \alpha_1^* \, [H+]_o \, ; \quad \alpha_{24} = \alpha_1^* \, [H^+]_{in}$$

$$\alpha_{35} = \alpha_2^* \, [L]_o \, ; \quad \alpha_{46} = \alpha_2^* \, [L]_{in} \qquad [7.3]$$

where α_{13}, α_{35}, α_{24}, α_{46} are the rate parameters for transitions 1 - 3, 3 - 5, 2 - 4 and 4 - 6, and α_1^*, α_2^* are concentration-independent coefficients. The kinetic equations [7.4] for the system are:

$$\frac{dN_1}{dt} = \alpha_{31} N_3 - \alpha_{13}N_1 + \alpha_{21}N_2 - \alpha_{12}N_1$$

$$[7.4]$$

$$\frac{dN_2}{dt} = \alpha_{42}N_4 - \alpha_{24}N_2 + \alpha_{12}N_1 - \alpha_{21}N_2$$

$$\frac{dN_3}{dt} = \alpha_{13} N_1 - \alpha_{31} N_3 + \alpha_{43} N_4 - \alpha_{34} N_3 + \alpha_{53} N_5 - \alpha_{35} N_3$$

$$\frac{dN_4}{dt} = \alpha_{24}N_2 - \alpha_{42}N_4 + \alpha_{34}N_3 - \alpha_{43}N_4 + \alpha_{64}N_6 - \alpha_{46}N_4$$

$$\frac{dN_5}{dt} = \alpha_{65}N_6 - \alpha_{56}N_5 + \alpha_{35}N_3 - \alpha_{53}N_5$$

$$N_1 + N_2 + N_3 + N_4 + N_5 + N_6 = N$$

By a choice of the constants α_{ij}, these differential equations may be solved by integration to determine the flux of lactose and H^+ through the membrane as functions of time.

Let us examine how this set of differential equations with many parameters can be reduced to a tractable analytical equation, using the known experimental and mechanistic details available from recent advances in studying lactose transport, especially from the work of Kaback's group (1979, 1980).

Basically, three kinds of experiments have been used by Kaback (1976) to study lactose translocation: exchange, influx and efflux. Experimentally, vesicles (intact cell membranes with the entire membrane machinery but without cytoplasmic constituents) are used as models for whole cells. During exchange, cells preloaded with 14C-lactose are diluted into equimolar substrate with unlabelled lactose, and the loss of labelled lactose is monitored. It should be noted that for exchange to occur the reaction sequences 4 - 6 - 5 - 3 and 4 - 5 - 6 - 4 can occur, whereas for influx or efflux the whole cycle must be completed. It is found experimentally that exchange is very fast, about 10 times the rate of influx or efflux. This indicates that the lactose section of the loop (4 - 6 - 5 - 3) is rapid compared to the H^+ part, and that the rate-determining step for efflux/influx must be in the steps 1 - 3, 1 - 2 or 2 - 4.

Assuming, then, that the lactose part of the loop is at equilibrium, we can derive rate expressions based on 1 - 3, 1 - 2 or 2 - 4 being rate-determining. It is also known that influx and efflux are affected by the ΔpH and $\Delta \psi$. Hence, any rate expression not involving H^+ concentrations or the equilibrium constant for sequence 1 - 2 (K_{12}) is eliminated. On this basis we arrive at the rate expressions shown in eqns. [7.5] and [7.6]. The ultimate value of these expressions depends, of course, on their validation by experimental data.

$$\text{Rate of influx} = \frac{dL_{in}}{dt} = \frac{A[t]\,L_o}{B[t] + L_o} = \frac{J_L}{V_c} \qquad [7.5]$$

$$A[t] = \frac{\alpha_{13}\,H_o^+\,N}{C[t]}$$

$$B[t] = \frac{K_{12}\,K'_{24}\,K'_{46}\,K_{65}\,K_{53}\,H^+_{in}\,L_{in}}{C[t]}$$

$$C [t] = 1 + K_{12} + K_{12} K'_{24} H^+_{in} + K_{12} K'_{24} H^+_{in} K'_{46} L_{in} + K_{12} K_{24} H^+_{in} K'_{46} L_{in} K_{65}$$

$$\text{Rate of efflux} = \frac{dL_{in}}{dt} = \frac{-A'[t]L_{in}}{B'[t] + L_{in}} \quad [7.6]$$

$$A'[t] = \frac{\alpha_{24} H^+_{in} N}{C'[t]}$$

$$B'[t] = \frac{K_{21} K'_{13} K'_{35} K_{56} K_{64} H^+_o L_o}{C'[t]}$$

$$C'[t] = 1 + K_{21} + K_{21} K'_{13} H^+_o + K_{21} K'_{13} K'_{35} H^+_o L_o + K_{56} K_{21} K'_{13} H^+_o L_o$$

where L_o = outside concentration of lactose,
  L_{in} = intracellular concentration of lactose,
  H^+_o = outside concentration of hydrogen ions,
  H^+_{in} = intracellular concentration of hydrogen ions,
  K_{ij}, K'_{ij} = are <u>time-dependent</u> coefficients, <u>in batch culture only</u>,
  A, B, C, A', B', C' = <u>time-dependent</u> parameters, <u>in batch culture only</u>,
  J_L = lactose flux, moles/sec.g,
  V_c = specific volume of cells, liters/g.

Table 7.3   Experimental Manipulations to Study Lactose Translocation (Kaczorowski et al., 1979)

Influx			
1 case	$\triangle \psi = 0$	$\triangle pH = 0$	facilitated diffusion,base
2	$\triangle \psi = -60mV$	$\triangle pH = -120mV$	D-lactate oxidation
3	$\triangle \psi = -60mV$	$\triangle pH = 0$	interior negative only
4	$\triangle \psi = 0$	$\triangle pH = -120$ mV	interior alkaline only
Efflux			
1	$\triangle \psi = 0$	$\triangle pH = 0$	base case
2	$\triangle \psi$		interior negative
3	$\triangle \psi$		interior positive
4	$\triangle pH$		interior alkaline
5	$\triangle pH$		interior acid

Table 7.3 shows a number of experimental manipulations on vesicles studied by Kaback's group to measure the kinetics of efflux and influx. For all these cases, the mathematical expressions given in eqns. [7.5] and [7.6] for the rates of influx and efflux predict the right trend of variation. Consider one illustrative example: for the 'normal' case of D-lactate oxidation (Kaczorowski and Kaback, 1979), $\triangle \psi$ = - 60mV, $\triangle$pH = - 120mV. For this base case, since the outside pH is buffered, the interior will be alkaline and hence $(H^+)_{in}$ will be low. Also, since the carrier is negatively charged, a negative membrane potential shifts equilibrium 1 - 2 toward 1 in the kinetic diagram (Fig. 7.4). Under these circumstances $A(t)$ is 115 nmol (mg protein)$^{-1}$ min^{-1}. Now consider facilitated diffusion where $\triangle \psi$ = 0, $\triangle$pH = 0. The interior is less alkaline compared to the reference case; i.e., $(H^+)_{in}$ increases, and $A(t)$ should decrease. This is precisely what is observed experimentally, $A(t)$ being 53 nmol (mg protein)$^{-1}$ min^{-1} for this case. Similarly, for all the cases in Table 7.3, the equations show the right trends.

7.5 MODEL VERIFICATION AND DISCUSSION

Quantitative verification of the model is important in establishing its validity. The approach taken toward verification in our laboratory involves comparison of model predictions with experimental data on two levels. First, experimental data from the literature were used to validate the transport model in an idealized system where transport is studied in isolation. Then, data from our experiments on the induction of β-galactosidase in *E. coli* were compared with model results for a range of cases which would occur in typical fermentations.

TRANSPORT OF LACTOSE (NO ENZYME INDUCTION)

Figures 7.5, 7.6, 7.7 and 7.8 demonstrate that the kinetic expressions derived for transport based on our mechanistic formulation agree with the experimental observations of Kaczorowski et al. (1979). The data were obtained on vesicles preloaded with labelled lactose under carefully defined conditions. This idealized system is a good candidate to ascertain whether eqns. [7.5] and [7.6] predict the solute translocation qualitatively. Note that the objective is not to construct a model to describe Kaczorowski's data, per se, but to corroborate the transport model which will be used (and verified) in a unified model for enzyme biosynthesis.   Figure 7.5 shows the efflux of lactose under an imposed negative electrical membrane potential.   The solid line represents the behavior predicted by eqn. [7.6], and the discrete points are experimental data from Kaczorowski et al. (1979).   Figure 7.6 corresponds to the data where a positive membrane potential was imposed on the vesicles.   Figures 7.7 and 7.8 show the effect of negative and positive Δ pH on lactose efflux from the vesicles.   In all four cases, agreement of the integrated simulation curves (solid lines) with the experimental data (Kaczorowski et al., 1979) is good, indicating that the transport model describes the behavior of these idealized cases, and may therefore be incorporated into a unified model for enzyme biosynthesis.

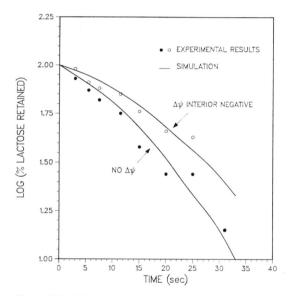

Figure 7.5   Efflux of lactose under a negative electrical gradient; $\circ$, $\bullet$ are experimental points obtained for lactose efflux from E. coli vesicles by Kaczorowski et al. (1979), $\bullet$ representing experimental conditions where the electrical gradient across the membrane was zero, while $\circ$ represents the imposition of an electrical membrane potential gradient (interior negative). The solid lines are solutions of the mathematical model by numerical integration of the differential equation by the Runge-Kutta 4 method (Δt=0.1s).

   Initial parameter values: (no $\Delta \psi$):   A=100.0 nmol (mg protein)$^{-1}$ min^{-1};   B=7.0 mM. ($\Delta \psi$) (interior negative):   A=50.0 nmol (mg protein)$^{-1}$ min^{-1};   B =3.0 mM.

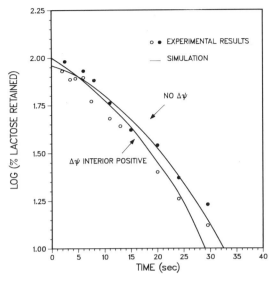

Figure 7.6   Efflux of lactose under a positive electrical gradient; O , ● are experimental points obtained for lactose efflux from *E. coli*   vesicles by Kaczorowski et al. (1979), ● representing experimental conditions where the electrical gradient across the membrane was zero, while *O* represents the imposition of an electrical membrane potential gradient (interior positive). The solid lines are solutions of the mathematical model by numerical integration of the differential equation by the Runge-Kutta 4 method ($\Delta t=0.1$s).

Initial parameter values: (no $\Delta \psi$):   A=100.0 nmol (mg protein)$^{-1}$ min^{-1}; B=7.0 mM. ($\Delta \psi$) (interior positive):   A=110.0 nmol (mg protein)$^{-1}$ min^{-1}; B=5.0 mM.

7.6 IMPLICIT INCLUSION OF $\Delta \psi$ IN THE TRANSPORT MODEL

To reiterate, it is believed that lactose is transported across the plasma membrane by co-transport with protons, coupled to the electrochemical gradient of protons across the membrane.   The total electrochemical gradient is comprised of the electrical gradient across the membrane ($\Delta \psi$) and the pH gradient (ΔpH). The rates of influx and efflux of lactose are affected by both $\Delta \psi$ and ΔpH. Equations [7.5] and [7.6] quantitatively describe the kinetics of influx and efflux. Note that each of the two processes describes a *net* rate of transport of lactose in an inward or outward direction; i.e., since the process of influx/efflux is cyclic, some lactose molecules will be transported inwards during efflux and outwards during influx. Equations [7.5] and [7.6] describe

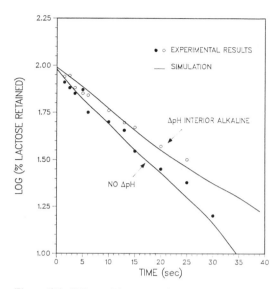

Figure 7.7   Efflux of lactose under a pH gradient (interior alkaline); O, ● are experimental points obtained by Kaczorowski et al. (1979) for lactose efflux from *E. coli* vesicles, ● representing experimental conditions where the pH gradient across the membrane was zero, while ○ represents the imposition of a pH gradient (interior alkaline) across the plasma membrane. Solid lines are solutions of the mathematical model by numerical integration of the differential equation by the Runge-Kutta 4 method ($\Delta t = 0.1 s$).

Initial parameter values: (no ΔpH):   A=180.0 nmol (mg protein)$^{-1}$ min^{-1}; B=8.0 mM. (Interior alkaline):   A=120.0 nmol (mg protein)$^{-1}$ min^{-1}; B=3.0 mM.

the net process.   It is immediately obvious from eqns. [7.5] and 7.6] that the effect of ΔpH on the transport kinetics appears explicitly as H^+_{in} and H^+_o.   The effect of $\Delta \psi$ on the transport is accounted for in an implicit fashion - by appropriate choice of K_{21} and K_{12} in the simulation.   Consider the schematic diagram of the kinetic sequence for lactose translocation (Fig. 7.4).   The only charged species involved in transport is the negatively charged carrier.   Therefore, any variation in the membrane electrical potential affects the transition 1 - 2 or 2 - 1. The normal state of the electrical potential is interior negative in an *E. coli* cell;   for D-lactate oxidation $\Delta \psi = -60mV$.   This implies that the interior of the cell membrane will be negatively charged and will repel the carrier towards the outside surface.   This is expected to cause K_{12} to be small.   For *influx*, since K_{12} appears in the denominator of the expression for A(t), A(t) will be large.   This gives an indication of the values of the parameters to be used in the simulation.   For influx of

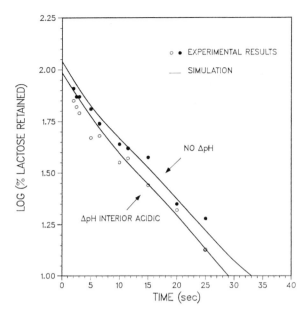

Figure 7.8 Efflux of lactose under a pH gradient (interior acidic); $\mathbf{O}$, $\bullet$ are experimental points obtained by Kaczorowski et al. (1979) for lactose efflux from *E. coli* vesicles, $\mathbf{O}$ representing the imposition of a pH gradient (interior acidic) across the plasma membrane. Solid lines are solutions of the mathematical model by numerical integration of the differential equation by the Runge-Kutta 4 method ($\Delta t = 0.1s$).

Initial parameter values: (no ΔpH): A=180.0 nmol (mg protein)$^{-1}$ min^{-1}; B=8.0 mM. (Interior acidic): A=120 nmol (mg protein)$^{-1}$ min^{-1}; B=3.0 mM.

lactose under these conditions, $A(t) = 115$ nmol/mg protein/min. When $\Delta \psi = 0$, the interior does not repel the carrier and K_{12} increases, causing a reduction in A; $A(t)$ drops to a value of 53 nmol/mg protein/min. For efflux (as is shown in Figs. 7.5 and 7.6), $\Delta \psi$ being negative implies that K_{21} will be large, causing a reduction in value of A. For Fig. 7.5, $A = 100$ ($\Delta \psi = 0$) and $A = 50$ ($\Delta \psi$ negative). This is because when $\Delta \psi = 0$, K_{21} becomes smaller and this being in the denominator of the expression for A, the value of A becomes larger. When $\Delta \psi$ is made interior positive (Fig. 7.6), the interior of the membrane actually attracts the negatively charged carrier; K_{21} will be even smaller. The effect of positive $\Delta \psi$ will be a higher value of A. This is what is found ($A = 110$ in Fig. 7.6). The effect of $\Delta \psi$ on the transport model therefore causes a change in the values of the parameters and is implicitly accounted for in the model.

7.7 INDUCER TRANSPORT IN FERMENTATION

Our experiments on the biosynthesis of β-galactosidase in *E. coli* involved fermentations with glucose, lactose or glucose-lactose mixtures in bench-scale fermenters with a nominal volume of 5 liters. The bacterial strain used in all the experiments was supplied by Professor Lloyd E. McDaniel, Waksman Institute of Microbiology, Rutgers University. It has been characterized as a mutant of *Escherichia coli* B (Strain PP01). Induction of the enzyme was achieved using lactose as the inducer or isopropylthio-D-galactoside (IPTG) as the 'gratuitous' inducer. Since the scope of this section is limited to inducer transport, verification of the unified set of model equations is not discussed just now in depth.

Briefly, the procedure for validation involved integration of the system equations to obtain profiles of the variables which are measured in the experimental fermentation runs. A nonlinear parameter estimation program was used to estimate values for the parameters and perform statistical tests to compare model predictions with experimental data. The tests indicated that the model results fit the experimental data quite well; the correlation coefficients were generally between 0.93 to 0.99.

Figure 7.9 illustrates the simulation and experimental results for growth and substrate depletion when the cells are grown on lactose; lactose also acts as an inducer for enzyme production. Fig. 7.10 illustrates a comparison of the simulation results for enzyme activity with the experimental data. The activity data show experimental scatter which is typical for this fermentation. The model, which includes transport of inducer, simulates the observed activity data reasonably well (correlation coefficient = 0.93). To verify whether the transport equation played a significant role, the nonlinear estimation program was run with and without the transport model. (This test was performed for all the fermentations on lactose and mixed substrates.) When the transport equation was included in the model, convergence of the nonlinear estimation was good and, for a number of initial parameter vectors, the estimation resulted in the same global parameter minimum.When either of the parameters (A or B) in the transport model was set equal to zero or when the transport model was eliminated, the parameter estimations were incapable of convergence to a realistic solu-

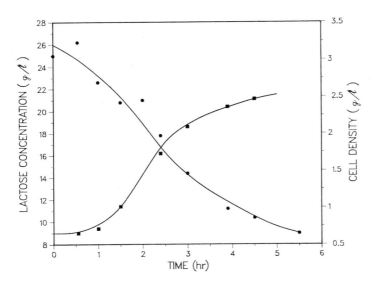

Figure 7.9   Cell growth and substrate depletion on lactose.   The points ● and ■ depict the experimental data for cell and lactose concentrations obtained during the fermentation of *E. coli* (PP01) grown on lactose as the sole carbon source.   Solid lines are simulation curves which represent the solution of the system equation.

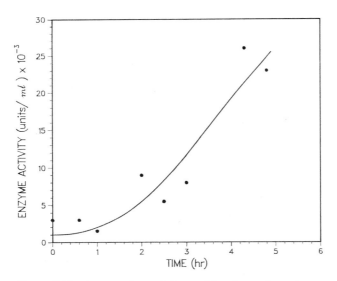

Figure 7.10   A comparison of the model and experimental enzyme activity profiles for induction by lactose.   The points ● depict the experimental enzyme activity data while the solid line is the simulation curve which represents the solution of the system equations.

tion.   It was reported in an earlier publication (Kaushik et al., 1979) that a transport equation was essential to simulate the production of glucose isomerase.   This requirement is to be expected, since the effective concentration of inducer responsible for regulation is the intracellular concentration.   Hence, it is erroneous to use the bulk concentration as the effective inducer concentration.   As we have seen, the internal and external concentrations are related by a nonlinear equation.

An interesting experimental variation is suggested by the induction experiments on pure lactose.   It is evident from the transport model that if the transported substrate is not consumed in the cell, it will accumulate and induce a high rate of enzyme biosynthesis.   IPTG is a nonmetabolizable (gratuitous) inducer and the introduction of IPTG into a batch culture of glucose-grown cells gives one important insights into the behavior of the model.

Table 7.4   Response to IPTG (threshold level)[a]

Experiment No.	Fermentation time (hr)	Cell concentration (g/l)	Activity (units)
1	16.0	3.30	0.0
2	19.0	3.23	0.0
3	20.0	--	0.0
4	22.5	--	0.0
5	24.5	--	0.0
6	26.0	--	0.0
7	40.0	3.23	0.0

[a]Initial glucose concentration, 15 g/l; IPTG concentration, 0.01443 g/l (6.057×10^{-5} M); IPTG introduced at 18.5 hr after start of run.

It is possible to determine the threshold level at which IPTG just causes induction of β-galactosidase activity.   Table 7.4 shows a low level of IPTG introduced into a fermentation carried out with glucose as the carbon source.   At this threshold level, no induction is observed.   Table 7.5 shows induction in a similar glucose-grown culture subjected to a feed of IPTG at about six times the threshold level.

Note that the threshold level (6 x 10^{-2} mM) is about 600 times below the normal inducing level of lactose used in the fermentation (ca. 35 mM). It is known (Kotyk and Janacek, 1970) that the accumulation ratio of IPTG is 10 to 100 times the bulk concentration, and that the *in vitro* potency of induction for IPTG and lactose are not far different. Now, if the model equations are solved using a hypothetical lactose concentration, using the same constants for transport as for the lactose case, the equivalent lactose 'threshold concentration' which causes no induction can be calculated. In addition, the 'equivalent lactose concentration' which generates the same profile as is observed in Table 7.5 can also be estimated.

Table 7.5   Response to IPTG (fed batch)[a]

Experiment No.	Fermentation time (hr)	Cell concentra- tion (g/l)	Enzyme activity (units)	Glucose concentra- tion (g/l)
1	2.25	0.281	0	15.0
2	3.25	0.281	0	16.6
3	5.25	0.265	0	17.0
4	7.25	0.265	0	--
5	10.25	0.273	0	16.6
6	24.25	2.63	--	--
7	25.75	3.27	622	--
8	27.0	3.74	1117	5.0
9	28.5	4.06	1531	0.0
10	30.0	4.14	2433	--
11	31.75	4.14	3566	--
12	33.75	4.10	3601	--

[a]Initial glucose concentration, 15 g/l; IPTG concentration, 0.0919 g/l (38.565×10^{-5} M); IPTG introduced at 24.25 hr after start of run.

If the transport constants are correctly obtained and if the model is valid, the equivalent threshold level of lactose should be ca. 1 - 5 mM. This same result is, indeed, predicted by the model. In addition, the 'equivalent lactose concentration' for the induction profile shown in Table 7.5 is ca. 15 mM. Figure 7.11 shows the model results with an 'equivalent lactose concentration' of 15 mM compared to the experimental enzyme activity data induced by IPTG at a concentration of 38 x 10^{-2} mM.   This corresponds to an accumulation ratio of 40, corroborating the value quoted earlier (10 - 100) (Kotyk and Janacek, 1970) and adding credibility to our transport model.

Basically, then, the model for transport was validated by comparing simulated results with the experimental data on lactose transport (in isolation) and in consort with the other mechanisms of enzyme induction (where it was found to be essential to predict the correct enzyme profiles).

These highlights of the verification of the simple, two-parameter, Monod-like transport model indicate that it accurately predicts the translocation of lactose in vesicles and also adequately determines the intracellular inducer level for enzyme induction in fermentation.   These findings were significantly extended in subsequent research in this area which is summarized below.

7.8 GENETIC REGULATION
OVERVIEW

Expression of the lactose (*lac*) operon in the *Escherichia coli* chromosome was studied in mixed-sugar chemostat cultures under

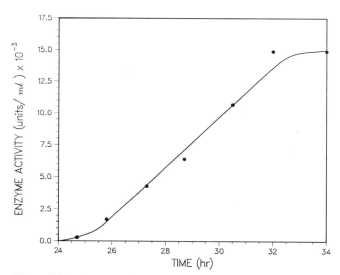

Figure 7.11   A comparison of the model results and the experimental enzyme activity profiles for induction by IPTG.   The solid line represents the model predictions using an 'equivalent lactose concentration' of 15 mM.   The points   ● depict the experimental enzyme activity data on glucose-grown cells which are induced by an IPTG concentration of 38x10^{-2} mM.

steady state and transient conditions. In what follows, a unified model is formulated which involves regulation of active inducer (lactose) transport, promoter-operator regulated expression of the lac operon, glucose-mediated inducer exclusion and catabolite repression. The model of the lac operon control system focuses on the molecular interactions among the regulatory species and the genetic control elements for the initiation of transcription. The role of catabolite modulator factor (CMF) in the regulation of transcription is described. The modeling of glucose-mediated regulation of intracellular cyclic adenosine monophosphate (cAMP) and inducer exclusion is based on the recently elucidated mechanisms of the involvement of the PTS (phosphoenolpyruvate dependent sugar transport system) enzymes, in the presence of glucose, in regulation of adenylate cyclase and non-PTS sugar transport proteins (i.e., permeases). The adequacy of the unified model was verified with experimental data.

Synthesis of lac operon enzymes is controlled at the transcription level. Any alterations in the regulatory elements may change the efficiency of the lac operon expression. Genetic alterations may occur in the recognition sites of the cAMP-CRP complex and RNA polymerase or in their binding sites or may be in the transcription starting point. Depending on the sites of alterations, the interactions of the regulatory species with the promoter region will be affected. For example, a mutation in the cAMP-CRP recognition site affects the binding affinity of cAMP-CRP with the DNA. This may also influence RNA polymerase binding to its recognition site, as with the model of Lee and Bailey (1984).

The model of lac operon transcription is based on molecular interactions among the various regulatory species and the genetic control units. Therefore, the parameters which are influenced by any genetic change (e.g., molecular cloning) can be easily isolated. For example, an alteration in the nucleotide sequence of cAMP-CRP binding sites may change the equilibrium constant k_2. As a result, the parameters K_B, K_C and K_E in eqn. [7.25], which are related to k_2, will be affected and that will influence transcription efficiency, F. Similarly, other interactions in the control elements will bring about changes only in the model parameters which correspond to the interactions involving the altered sites. Furthermore, alterations in the control units of the genes which code for the regulatory proteins can result in changes in the rate of synthesis of those proteins which take part in the control of lac operon expression, as will be shown in Chapter 8.

INTRODUCTION

Microorganisms regulate their metabolic activities in response to changes in abiotic conditions. Most often the control system adjusts the regulatory interactions in order to efficiently utilize the energy resources for cellular growth. One such example is the diauxic growth of E. coli, when both glucose and lactose are present in the culture as the carbon sources. It has been observed that E. coli preferentially takes up and metabolizes glucose over lactose (Nakada and Magasanik, 1964). This ability of regulation, in a physiological sense, enables the bacteria to use their energy more effectively, since the specific proteins are synthesized only when needed.

Although the diauxic phenomenon is well known, the underlying molecular mechanisms which regulate the cell functions to preferentially metabolize one substrate over the other were incompletely known until recently (Nelson et al., 1983; Mitchell et al., 1982). The PTS proteins are known to be involved in glucose uptake from the external medium across the plasma membrane of E. coli (Postma and Roseman, 1976; Saier, 1977). Membrane-bound glucose specific enzyme II^{Glc} translocates glucose and cytoplasmic enzyme III^{Glc} transfers the phosphate group to glucose inside the cell. It has been indicated that the enzyme III^{Glc} of the PTS in its non-phosphorylated form inhibits transport of non-PTS sugars by regulating activities of their transport proteins (Nelson et al., 1983; Scholte et al., 1981). Therefore, when both glucose and lactose (non-PTS sugar in E. coli) are present in the medium, the presence of unphosphorylated III^{Glc}, as a result of PTS mediated glucose transport, causes repression of lactose permease and thus inhibits lactose translocation. Since lactose is a lac operon inducer, this mechanism is also known as "inducer exclusion." In addition to inducer exclusion, the non-phosphorylated III^{Glc} is also believed to be involved in the regulation of adenylate cyclase, which catalyzes the synthesis of cAMP from ATP (Postma et al., 1981; Dills et al., 1980). cAMP is known to be a modulator of catabolite repression (Pastan and Adhya, 1976; Epstein et al., 1975). The presence of glucose or other readily metabolizable substrates in the medium exerts a permanent repression of catabolic enzymes. The glucose regulated repression is thought to be caused by reduced cAMP levels in the cells (Makman and Sutherland, 1965). Subsequent studies demonstrated that cAMP , via cAMP receptor protein (CRP), positively modulates the transcription of a large number of genes in E. coli (Zubay, 1980). cAMP and CRP form a genetic regulatory complex which binds to specific nucleotide sequences in the promoter regions. This interaction is required for the RNA polymerase to bind to its specific sites on the promoter and thereby promote transcription.

The role of cAMP-CRP complex as a positive regulator to relieve catabolite repression led to the paradigm that the extent of catabolite repression is strictly controlled by the intracellular concentration of cAMP. However, increasing evidence based on recent studies suggests that cAMP-CRP complex may not be the exclusive regulator of catabolite repression (Lee and Dobrogosz, 1983; Guidi-Rontani et al., 1980; Joseph et al., 1982). Ullmann et al. (1976) found that water soluble extracts of *E. coli* caused strong repression of the expression of catabolite sensitive operons. They have partially purified the compound responsible for this effect and have designated it as catabolite modulator factor (CMF). The catabolite modulator factor, the mediator of negative control, is believed to interact with a specific RNA polymerase factor and inhibit its activity to carry out transcription (Ullman and Danchin, 1983). cAMP-CRP complex, however, counteracts the action and partially overcomes the repressive effects (Dessein et al., 1978). The accumulation of CMF from active metabolism of glucose is thought to be dependent on growth conditions. It has been postulated that the growth conditions which would favor the accumulation of catabolites, would also enhance the rate of synthesis of the modulator.

There have been several studies on quantitative modeling of inducible enzyme biosynthesis, based upon induction-repression mechanisms (Yagil and Yagil, 1971; Venkatasubramanian et al., 1976; Toda, 1976; Imanaka and Aiba, 1977; Gondo et al., 1978; Lee and Bailey, 1984). Lee and Bailey also described the expression of the lac operon cloned in multicopy plasmids. Some of the papers which dealt with mixed sugar fermentation incorporated the role of cAMP as the exclusive modulator of catabolite repression. Vieth et al. (1982) incorporated the inducer transport model, describing active translocation of lactose across the plasma membrane, with the enzyme biosynthesis model. None of these models, however, dealt with the role of CMF on catabolite repression or the role of the PTS enzyme III^{Glc} in the regulation of intracellular cAMP and inducer transport.

7.9 MODEL ELEMENTS
GROWTH ON MIXED SUBSTRATE

In batch culture experiments, with glucose or lactose as carbon source, a progressive decrease in specific growth rate has been observed in the growth phase. A product inhibition model was employed in simulations of culture growth.   Competitive inhibition by one substrate to the consumption of the other was also included:

$$\frac{dX_1}{dt} = \frac{\mu_{m1} S_1 X_1}{S_1 + K_{S1} [1 + S_2 /K_{I2} + \sum_{j=1}^{2} P_{ij} /k_{ij}]} \qquad [7.7]$$

$$\frac{dX_2}{dt} = \frac{\mu_{m2} S_2 X_2}{S_2 + K_{S2} [1 + S_1 /K_{I1} + \sum_{j=1}^{2} P_{ij} /k_{ij}]} \qquad [7.8]$$

$$X = X_1 + X_2 \qquad [7.9]$$

where X is the cell density in the culture, S is residual sugar concentration, K_S represents the Monod saturation constant, μ_m is the maximum specific cell growth rate, P_i is the concentration of growth inhibitory product, and K_I and k_i represent the coefficients of inhibition.   Subscripts 1 and 2 indicate correspondence with glucose and lactose, respectively.   In a steady state chemostat culture, for "growth-linked" inhibitory product:

$$P_i = Y_{P_i} [S_o - \widetilde{S}] \qquad [7.10]$$

where Y_{P_i} = product (inhibitory) yield coefficient. S_o and $\widetilde{S}$ are the inlet and steady state sugar concentrations, respectively.   Let,

$$K_P = \frac{Y_{P_i}}{k_i} \qquad [7.11]$$

The equations for substrate consumption are:

$$- r_{S_1} = \frac{1}{Y_{X_2}} r_{X_1} \qquad [7.12]$$

$$- r_{S_2} = \frac{1}{Y_{X_2}} \, r_{X_2} \qquad\qquad [7.13]$$

where Y_X is a growth yield coefficient.

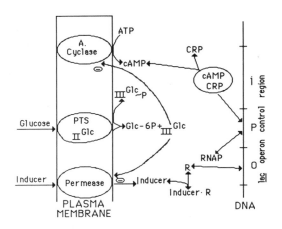

Figure 7.12   Schematic diagram of the regulation of lac operon transcription in mixed-sugar (glucose + lactose) fermentation.

GENE EXPRESSION KINETICS

The present model of the expression of the catabolite sensitive lac operon accounts for the molecular interactions among the regulatory species and lac genetic control units at the initiation of transcription. It also incorporates the transport model to account for inducer translocation across the plasma membrane.   Furthermore, the model deals with the involvement of the PTS enzymes in regulating intracellular cAMP and inducer uptake.   Figure 7.12 shows a schematic diagram of the framework for the present model.

The rate of synthesis of the lac mRNA by transcription and the rate of formation of β-galactosidase by translation may be described by the following equations:

$$\frac{d[mRNA]}{dt} = K_{+M} \, F \, Q \, \mu - K_{-M} \, [mRNA] - \mu \, [mRNA] \qquad [7.14]$$

$$\frac{d[E]}{dt} = K_{+E} \, [mRNA] - K_{-E} \, [E] - \mu \, [E] \qquad [7.15]$$

where (mRNA) and (E) represent intracellular concentrations of mRNA and β-galactosidase, respectively. FQ is the overall transcription efficiency. F represents efficiency of the initiation of transcription at the promoter and Q represents induction efficiency at the operator site. It has been assumed that the decay of mRNA and E follow first order kinetics. The last terms of eqns. [7.14] and [7.15] take into account the effect of dilution due to cell growth.

The Q function, the fraction of repressor-free operator, has been shown to be related to the intracellular inducer concentration according to the following equation (Venkatasubramanian et al., 1976):

$$Q = \frac{1 + b_1 \, I_{in}}{1 + b_1 \, I_{in} + b_2} \qquad [7.16]$$

where, I_{in} is the intracellular inducer concentration, b_1 is a constant and $b_2 = b'_2 \, (R)$. (R) is the concentration of lac repressor protein and b_2 is a constant.

7.10 INDUCER TRANSPORT IN CHEMOSTAT CULTURE

The intracellular concentration of lactose, inducer of the lac operon, has been related to the external concentration by the lactose transport model (Vieth et al., 1982). The rate expression in simplified lumped form can be expressed as:

$$\frac{dL_{in}}{dt} = \frac{A'' \, [t] \, L_B \, \mu}{B \, [t] + L_B} - \mu \, L_{in} \qquad [7.17]$$

where $A''(t) = A(t)/\mu$ and L_{in} and L_B are intracellular and bulk lactose concentration, respectively. $A''(t)$ and $B(t)$ are weakly time-dependent transport coefficients in transient culture, while they are true constants in continuous culture.

7.11 CATABOLITE REPRESSION

The model developed in this section describes the extent of catabolite repression in terms of efficiency of initiation of transcription at the promoter site of the lac operon. Shown earlier is a schematic diagram of the transcriptional controls at the lac promoter-operator in Figure 7.12. Earlier models in this area were based on the paradigm that the modulation of the extent of catabolite repression was exclusively controlled by the intracellular concentration of cAMP. However, the present model accounts for an additional control mechanism at the transcriptional level.

The existence of the CMF regulation in the expression of catabolite sensitive operons leads to the development of the present model which accounts for the extent of catabolite repression in terms of a balance between the negative control by CMF and the positive regulation by cAMP-CRP at the initiation of transcription. In developing the model of the lac operon control system in the promoter region, the following assumptions have been made.

(1)   A single copy of the catabolite sensitive lac operon is present in the chromosomal DNA and the expression of this operon is subject to genetic regulations under the control of lac promoter-operator in a wild-type strain of *E. coli* .

(2)   The binding of the cAMP-CRP (CC) to the specific site of the lac promoter (P) makes an intermediate complex (INTC). The formation of the complex destabilizes the DNA helix, thus facilitating the RNA polymerase (RNAP) to interact with the specific sequence of the promoter and to form an "open" complex (OPN 1) which initiates transcription at a uniform rate.

(3)   A direct RNAP-P interaction in the absence of cAMP-CRP results in a shift of the positioning of the RNAP. Though this may form an "open" complex (OPN 2), it is unstable and mRNA transcription is initiated at a much reduced rate.

(4)   The catabolite modulator factor (M), the mediator of negative control, interacts with a specific RNA polymerase factor and inhibits its activity to bind to the promoter region. However, cAMP-CRP complex reduces the extent of catabolite repression by stimulating the RNAP-P interaction. No previous modeling efforts have considered this role of CMF in the regulation of gene expression quantitatively.

The equilibrium interactions among the regulatory species participating in the initiation of transcription may be written as follows:

$$cAMP + CRP \overset{k_1}{\Leftrightarrow} C\text{-}C$$

$$C\text{-}C + P \overset{k_2}{\Leftrightarrow} INTC$$

$$INTC + RNAP \overset{k_3}{\Leftrightarrow} OPNI$$

$$P + RNAP \overset{k_4}{\Leftrightarrow} OPN\ II$$

$$M + RNAP \overset{k_5}{\Leftrightarrow} RNAP\text{·}M$$

where k_1 through k_5 are the equilibrium constants, given by:

$$k_1 = \frac{[C\text{-}C]}{[cAMP]\,[CRP]} \qquad [7.18]$$

$$k_2 = \frac{[INTC]}{[C\text{-}C]\,[P]} \qquad [7.19]$$

$$k_3 = \frac{[OPNI]}{[INTC]\,[RNAP]} \qquad [7.20]$$

$$k_4 = \frac{[OPN\ II]}{[P]\,[RNAP]} \qquad [7.21]$$

$$k_5 = \frac{[RNAP \cdot M]}{[M]\,[RNAP]} \qquad [7.22]$$

defining, $$f = \frac{[OPNI] + [OPN\ II]}{[P]_t} \qquad [7.23]$$

where, $[P]_t = [P] + [INTC] + [OPNI] + [OPN\ II]$ [7.24]

We arrive at the following relationship to define the F function:

$$F = \frac{f}{f_{max}} = \frac{K_A + K_B\ \gamma}{1+K_A+K_C\ \gamma+[M]\ [K_D+K_E\ \gamma]} \cdot \frac{1+K_A+K_C}{K_A + K_B}$$ [7.25]

where, $K_A = k_4\ [RNAP]_t$ [7.26]

$K_B = k_1\, k_2\, k_3\ [CRP]_t\, [RNAP]_t + k_1\, K_A\ [cAMP]_{max}$ [7.27]

$K_C = (k_1 + k_1\, k_2\ [CRP]_t)\ [cAMP]_{max} + K_B$ [7.28]

$K_D = k_5$ [7.29]

$K_E = k_1\, k_5\ (1 + k_2\ [CRP]_t)\ [cAMP]_{max}$ [7.30]

$$\gamma = \frac{[cAMP]}{[cAMP]_{max}}$$ [7.31]

$[RNAP]_t = [RNAP] + [RNAP \cdot M]$ [7.32]

$[CRP]_t = [CRP] + [C\text{-}C]$ [7.33]

In developing eqn. [7.25], it is assumed that RNAP and CRP molecules bound to the promoter region are in much lower concentration compared to those in the nonbound form. If (M) is zero, eqn. [7.25] becomes similar to an expression described by Gondo et al. (1978).

7.12 KINETICS OF CMF SYNTHESIS

As discussed, CMF exerts negative control at the initiation of transcription for the expression of catabolite sensitive lac operons. As shown in eqn. [7.25], intracellular accumulation of this compound reduces the value of F. The magnitude of catabolite repression and the extent of its relief by cAMP-CRP complex is dependent to a large extent on the intracellular concentration of the mediator, which, in turn, is dependent on its rates of synthesis and degradation. CMF is found to be actively metabolized by the cells (Dessein et al., 1978). Since CMF is an intermediary metabolite and is produced by glucose metabolism, it can be construed that the synthesis of CMF is linked to the substrate (glucose) utilization rate. If it is assumed that the degradation (metabolism) of the mediator follows first order kinetics, then the rate expression becomes:

$$\frac{d[M]}{dt} = k_m \frac{\mu}{Y_x} - k_{-m} [M] - \mu[M] \qquad [7.34]$$

where,

$$\frac{\mu}{Y_x} = \text{specific substrate utilization rate,}$$

k_m is a constant and k_{-m} is the decay constant.

In a continuous culture under steady state conditions $\dfrac{d[M]}{dt} = 0$

and $\mu = D$. Therefore,

$$[M] = \frac{[k_m / Y_x] D}{k_{-m} + D} \qquad [7.35]$$

It has been observed on many occasions that the extent of catabolite repression during inducible enzyme biosynthesis is directly related to the specific cell growth rate, even when a substrate inducer is present as the only carbon source (Schaffer and Cooney, 1982; Clarke et al., 1968). In chemostat experiments it is found that catabolite repression was high at high dilution rates, whereas, at low dilution rates, induction was dominant and catabolite repression was minimal. The general conclusion was that at high growth rates the repressive catabolite(s) were being formed at increased rates to cause significant repression. At low growth rates, however, the concentration of catabolites was low. Ullmann and Danchin (1983) postulated that the growth conditions which would lead to the accumulation of catabolites, would also favor the synthesis of CMF. Therefore, it can logically be concluded that at low growth rates the accumulation of CMF is not favored. It seems reasonable, therefore, on the basis of the above discussion, to assume a critical substrate utilization rate, below which the synthesis of CMF does not occur. In a batch culture this "below-critical" situation generally does not arise before the near-end of the exponential phase. However, in a substrate limited chemostat culture, specific growth rate and the specific substrate utilization rate can be controlled. If D_C represents the dilution rate corresponding to the critical substrate utilization rate, then eqn. [7.35] may be used, substituting $(D - D_C)$ for D , to calculate (M).

7.13 PTS MEDIATED REGULATION OF INTRACELLULAR CAMP AND LACTOSE TRANSPORT

PTS sugars are transported into bacteria by a group translocation mechanism which requires phosphorylation of the sugars. The sugars which are PTS substrates negatively regulate the induction of the necessary enzyme systems for the metabolism of non-PTS sugars. In *E. coli*, glucose specific enzyme III^{Glc} of the PTS in its unphosphorylated form is believed to be involved in the inhibition of adenylate cyclase activity, and also, in the regulation of non-PTS sugar (e.g., lactose in *E. coli*) permease activity. Therefore, the extent of PTS regulation is dependent on the intracellular concentration of III^{Glc}.

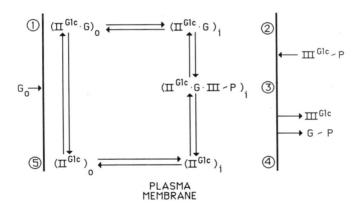

Figure 7.13   Schematic diagram of the group translocation via PTS. G_0 represents glucose concentration in the external medium.

KINETICS OF UNPHOSPHORYLATED IIIGLC FORMATION

A schematic representation of glucose transport is shown in Fig. 7.13. The membrane bound enzyme II^{Glc} catalyzes the transport across the plasma membrane. The transfer of phosphoryl group from $III^{Glc} \sim P$ to glucose is also catalyzed by II^{Glc}. The reaction sequence of the group translocation can be described as follows:

$$[G]_o + [II^{Glc}]_o \overset{K_1}{\Leftrightarrow} [II^{Glc} G]_o \overset{K_2}{\Leftrightarrow} [II^{Glc} G]_i$$

$$[II^{Glc} G]_i + [III^{Glc} \sim P] \overset{K_3}{\Leftrightarrow} [II^{Glc} G III^{Glc} \sim P]_i \overset{K_4}{\Leftrightarrow} [II^{Glc}]_i + [G \sim P] + [III^{Glc}]$$

Kinetics can be derived on the basis of the rate limiting step, which is considered to be the step 3 - 4. The other reaction steps are in equilibrium with constants, K_1, K_2 and K_3. Using the above relationships the rate equation yields:

$$\frac{d[III^{Glc}]}{dt} = \frac{K_R [G]_o}{K_{R S} + [G]_o} \mu - K_{RD} [III^{Glc}] - \mu [III^{Glc}] \qquad [7.36]$$

where,

$$K_R \mu = K_1 K_2 K_3 K_4 [II^{Glc}]_t \cdot [III^{Glc} \sim P] / [K_1 + K_1 K_2 + K_1 K_2 K_3 [III^{Glc} \sim P] \qquad [7.37]$$

$$K_{R S} = 1/[K_1 + K_1 K_2 + K_1 K_2 K_3 [III^{Glc} \sim P]] \qquad [7.38]$$

$$K_{RD} = \text{Decay constant}$$

$$[II^{Glc}]_t = [II^{Glc}] + [II^{Glc} G]_o + [II^{Glc} G]_i + [II^{Glc} G III^{Glc}] \qquad [7.39]$$

7.14 REGULATION OF ADENYLATE CYCLASE ACTIVITY

cAMP is synthesized from ATP, catalyzed by membrane-bound adenylate cyclase. The enzyme catalyzed reaction is shown below:

$$ATP \rightarrow cAMP + PP_i$$

where PP_i is pyrophosphate.

Unphosphorylated III^{Glc} regulates adenylate cyclase activity, leading to a decrease in the intracellular cAMP levels. The general sequence of reactions is as follows:

$$[E_{AC}] + [S] \overset{k_a}{\Leftrightarrow} [E_{AC} \cdot S] \overset{k'}{\Rightarrow} [E_{AC}] + [cAMP]$$

$$[E_{AC}] + [III^{Glc}] \overset{k_b}{\Leftrightarrow} [E_{AC} \cdot III^{Glc}]$$

$$[E_{AC} \cdot S] + [III^{Glc}] \overset{k_c}{\Leftrightarrow} [E_{AC} \cdot S \cdot III^{Glc}]$$

$$[E_{AC} \cdot III^{Glc}] + [S] \overset{k_e}{\Leftrightarrow} [E_{AC} \cdot S \cdot III^{Glc}]$$

where (E_{AC}), (S) and cAMP are intracellular concentration of adenylate cyclase, ATP and cAMP respectively. k_a, k_b, k_c and k_e are equilibrium constants. It is assumed that cAMP can emerge only from the reaction $[E_{AC} \cdot S] \Rightarrow (E_{AC})$ [cAMP], which is assumed to be the rate limiting step in the overall reaction. We finally arrive at a rate expression which has the form:

$$\frac{d[cAMP]}{dt} = \frac{k'_a \mu}{1+k'_b+k'_c (III^{Glc})} - k_d [cAMP] - \mu [cAMP] \qquad [7.40]$$

where,

$$k'_a \mu = k'k_a [E_{AC}]_t [S] \qquad [7.41]$$

$$k'_b = k_a [S] \qquad [7.42]$$

$$k'_c = k_b + k_a k_c [S] + k_b k_e [S] \qquad [7.43]$$

$$[E_{AC}]_t = [E_{AC}] + [E_{AC} \cdot S] + [E_{AC} \cdot III^{Glc}] + [E_{AC} \cdot S \cdot III^{Glc}] \qquad [7.44]$$

k_d is the rate constant for cAMP depletion. cAMP is depleted by excretion (Makman and Sutherland, 1965) and a phosphodiesterase catalyzed degradation reaction. However, it has been found that phosphodiesterase activity is weak in *E. coli* (Alper and Ames, 1975).

7.15 INDUCER EXCLUSION

One of the glucose mediated regulations to prevent induction of the non-PTS sugar enzymes is to inhibit the entry of the inducer into bacteria (inducer exclusion).   Available evidence suggests that glucose-specific enzyme III^{Glc} of the PTS in its non-phosphorylated form directly binds to the various non-PTS transport systems and thus inhibits their activities (Mitchell et al., 1982).   Saier et al. (1983) observed that the binding of III^{Glc} to the lac permease showed cooperativity in substrate binding.   Therefore, binding of lactose to its permease is required for effective regulation by the III^{Glc} .

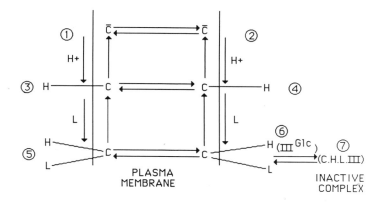

Figure 7.14   Schematic representation of inducer translocation and exclusion.   H$^+$, L and C represent proton, lactose and carrier, respectively.

In developing the mathematical description of inducer exclusion, inhibition of lac permease activity by III^{Glc} during active transport of lactose across the plasma membrane was considered as the model system.   The general sequence of kinetic steps involved in the lactose-proton symport is shown in Fig. 7.14.   It has been found that lactose exlcusion by PTS-III^{Glc} does not involve interruption of energy coupling to solute accumulation (Mitchell et al, 1982).   Since III^{Glc}, a cytoplasmic protein, interacts with the lac carrier only when lactose is bound to the carrier, it is likely that the CHL complex at the cytoplasmic site of the membrane would be the target of III^{Glc}. The schematic diagram of the lactose translocation-exclusion mechanism is shown in Fig. 7.14. For influx of lactose the net sequence is 1 - 3 - 5 - 6 - 4 - 2. The step 1 - 3 is considered to be the rate-determining step in the overall process of lactose influx. Assuming other binding sites are in equilibrium, we arrive at a rate expression which in simplified form is:

$$\frac{dL_{in}}{dt} = \frac{1}{1+B''[t]\,[III^{Glc}]} \cdot \frac{A''[t]\,L_B\,\mu}{\dfrac{B[t]}{[1+B''[t]\,[III^{Glc}]\,]} + L_B} - \mu L_{in} \qquad [7.45]$$

where $A''(t)$ and $B(t)$ are identical to the terms in eqn. [7.17].

$$B''[t] = \frac{B[t]\,K_{67}}{K_{53}\,K_{65}} \qquad [7.46]$$

where K_{53}, K_{65} and K_{67} are the equilibrium constants. At steady state,

$$L_{in} = \frac{1}{1+B''[t]\,[III^{Glc}]} \cdot \frac{A''[t]\,L_B}{\dfrac{B[t]}{[1+B''[t]\,[III^{Glc}]\,]} + L_B} \qquad [7.47]$$

The presence of (III^{Glc}) in eqn. [7.47] reduces the value of maximum intracellular lactose concentration by a factor of $1 / 1+B''(t)$ (III^{Glc}). This is quite expected, because the lactose carriers bound to III^{Glc} are inactive in transporting sugar. The change of saturation constant in eqn. [7.47] is also consistent with experience, since availability of fewer lac carriers would show saturation effects at lower bulk lactose concentration (L_B).

7.16 MODEL VERIFICATION

The model so developed can be used in a wide range of growth conditions for the expression of the lac operon in the presence of lactose, as well as a mixture of glucose and lactose as the carbon source(s). Model verification with lactose as the only carbon source had been reported earlier (Ray et al., 1986). The presence of glucose in the growth medium alters the functions of the lac operon control system, since glucose influences some of the regulatory species which are involved in transcriptional control and inducer translocation. Therefore, modeling lac operon functions becomes more complex with the growth medium containing both glucose and lactose. In a mixed-substrate batch fermentation, when glucose concentration in the culture is high, expression of the lac operon is primarily suppressed by inducer exclusion. Even if some inducer is present in the cytoplasm of a preinduced cell, the control of intracellular cAMP by glucose would severely inhibit the formation of "open" complex to start transcription. In a chemostat culture, however, the residual glucose concentration in the culture is generally below a critical level to completely shut off inducer entry and transcription. The steady state intracellular cAMP levels would depend on concentration of the unphosphorylated III^{Glc} which again is dependent on external glucose concentration, as indicated by eqn. [7.36]. Therefore, γ in eqn. [7.25] is a variable. The inducer transport-exclusion model, as described by eqn. [7.47], predicts the internal lactose concentration.

7.17 PARAMETER ESTIMATION

The model equations contain a large number of parameters. The parameter values shown in Table 7.6 were determined based on information in the literature and estimation from the experimental data using a nonlinear statistical program package available on the system library NONLIN (Metzler et al., 1974). The program uses an improved Gauss-Newton search technique and is capable of estimating parameters when the system equations are defined by algebraic as well as differential equations. Also, it is capable of performing statistical analyses of model correlation with experimental data.

Those values of the parameters are chosen which give the best fit of the model to the experimental data, in the least square sense. A number of criteria have been used to ensure the validity of our parameter estimation. The least square estimate is judged by comparing the correlation coefficient with the perfectly correlated value of unity. The nonlinear estimation program also evaluates 95% univariate confidence limits on the evaluated parameters for a normal distribution of errors using the t-statistic and the allowed degrees of freedom. If the confidence limits span both sides of zero, the parameter may be proven unnecessary in the model, because the parameter could be zero. Although the analysis is valid for both linear systems and for nonlinear models, the argument is only applicable for a piecewise linearized approximation of the complete equation set in the region of the optimized parameter array.   Many of the parameter values were evaluated from literature information (Ray et al., 1986).   The parameters to be estimated by the NONLIN program were further reduced by evaluating some of them stepwise by independent experiments.   Parameters which were found to be statistically insignificant were dropped.

The values of parameters K_{S_1} and K_{S_2} (in growth equations) for *E. coli* on glucose and lactose are reported by Pirt (1975). These values, which are 0.004 g/l and 0.02 g/l respectively, have been used in the simulation.   Other parameters in the growth equations were estimated by batch-and chemostat-culture experiments.   The transport constant A"(t) in eqn. [7.47] was determined directly from an independent experiment.   The half-life of mRNA has been reported to be about 1.5 min (Pichon et al., 1977). Based on this value K_{-M} has been estimated to be 27.6 hr^{-1}.   The reported values of K_{+E} and b_1 are 1000.0 units/(mg mRNA)(hr) and 5.0 l/g respectively (Vieth et al., 1982). These values have been used in the simulation. The lumped constants K_A, K_B and K_E in eqns. [7.26], [7.27] and [7.30], respectively, which are defined in terms of binding affinities and intracellular concentrations of regulatory proteins and effectors, were at first estimated (in an order-of-magnitude sense) from available information in the literature of molecular biology.   For instance, K_B is defined as:

$$K_B = [k_1 k_2 k_3 \, [CRP]_t \, [RNAP]_t + k_1 K_A] \, [cAMP]_{max}$$

where, $K_A = k_4 \, [RNAP]_t$

The reported values of k_1, k_2 and k_3 are approximately in the ranges 10^4 - 10^5 M^{-1} (Takahashi et al., 1980; Anderson et al., 1971), 10^7 - 10^8 M^{-1} (Takahashi et al., 1983) and 10^8 - 10^9 M^{-1} (Strauss et al., 1980), respectively.   It has been observed that in the absence of cAMP-CRP the open complex formed by RNA polymerase binding to the lac promoter is not stable (Spassky et al., 1984). Therefore, the k_4 value is much less than k_3.   The number of CRP molecules in a cell is generally > 1000 (Anderson et al., 1971).   The estimated numbers of intracellular core RNA polymerase and σ subunit molecules are approximately 2500-3000 and 700, respectively (Iwakura et al., 1974).   It has been estimated that one molecule in an *E. coli* cell represents approximately 10^{-9} M (Miller, 1972). The maximum values of intracellular cAMP in *E. coli* are reported to be in the range 5-10 μM (Pastan and Adhya, 1976). We have used $(cAMP)_{max} = 10\mu M$ in this work. On the basis of the above information we have used $K_A = 0.1$, $K_B = 60.0$ and $K_E = 3.0$ in the simulation. K_C was estimated as 68.0; k'_b of eqn. [7.42] is given by $k_a(S)$. The steady state level of intracellular ATP in *E. coli* is reported to be approximately 2mM (Wilson et al., 1980). If k' is assumed to be very small compared to the forward and reverse reaction rate constants of $(E_{AC}) + (S) \Leftrightarrow (E_{AC} \cdot S)$ then k_a can be approximated from the literature value to be 2.0×10^3 M^{-1} (Ullmann and Danchin, 1983). On that basis the value of K'_D was calculated to be 4.0.

The steady state intracellular cAMP concentration has been assumed constant during lactose fermentation in chemostat culture. Accumulation of unphosphorylated III^{Glc} during lactose fermentation is not implied, since transport of glucose from the external medium is necessary for this, although glucose is formed in the interior of the cells by lactose hydrolysis. Wright et al. (1979) found no significant change in intracellular cAMP in *E. coli* even when bacterial growth rates were varied eight-fold in carbon limited chemostat cultures. A value of $(cAMP)_{ss} = 5\mu M$ has been assumed (on lactose). Accordingly, a value of k'_a equal to 8.3 mg/l was used in the simulation. The maximum intracellular concentration of III^{Glc} has been reported to be approximately 1000 mg/l (Postma et al., 1981). Therefore, the value of K_R of eqn. [7.36] is assumed to be 1000 mg/l.

FAMILY OF PARAMETERS

Although the individual members of the array of experiments conducted in this work correspond to different environmental conditions, which have significant effects on the cellular physiology, the remaining constants estimated using NONLIN from different runs were found to be consistently in the same magnitude, which indicates that the model is sufficiently robust and general and can be applied in a wide range of conditions. Table 7.6 shows the parameter values for cell growth and β-galactosidase induction in a mixed-sugar chemostat culture.

7.18 STEADY STATE RESPONSE IN CHEMOSTAT CULTURES

Figures 7.15 and 7.16 compare the model predictions with experimental data for specific β-galactosidase activity in mixed-sugar

Table 7.6   List of Parameter Values

(Feed Sugar: Glucose 3.0 g/l and Lactose 3.0 g/l)

μ_{m1} = 0.98 hr^{-1} μ_{m2} = 0.92 hr^{-1}
K_{S1} = 0.004 g/l K_{S2} = 0.02 g/l
K_{p1} = 2.27 l/g K_{p2} = 5.0 l/g
K_{11} = 10^{-3} g/l K_{12} = 6.6 x 10^{-3} g/l
K_{+M} = 53.9 mg/g K_{-M} = 27.6 hr^{-1}
K_{+E} = 10^3 units/mg.hr b_1 = 5.0 l/g
b_2 = 100   A" = 60 g/l
B = 0.45 g/l K_A = 0.1
K_B = 60.0 K_C = 68.0
K_D = 1.0 l/mg K_E = 3.0 l/mg
k_m = 0.94 mg/l k_{-m} = 0.064 hr^{-1}
K_R = 10^3 mg/l K_{RS} = 0.8 g/l
k'_a = 8.3 mg/l k'_b = 4.0
k'_c = 0.1 mg/l   B" = 0.028 l/mg

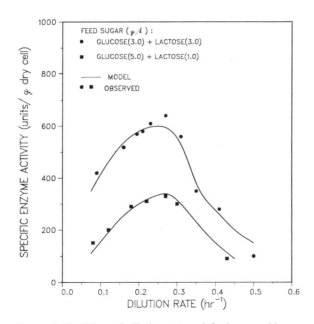

Figure 7.15   Effect of dilution rate and feed compositions on steady state enzyme specific activity in a chemostat culture (mixed sugar medium).

chemostat cultures.   Total feed sugar in Figs. 7.15 and 7.16 was 6.0 and
9.0 g/l, respectively.   Lactose to glucose ratio was varied and steady
state enzyme specific activity was measured as a functon of dilution
rate.   Specific activity was increased as the ratio was raised.   The rising
parts of the curves at low dilution rates are due to increased level of
induction because of higher intracellular inducer concentration as the
dilution rate is increased and the declining parts of the curves are due
to domination of catabolite repression and inducer exclusion over
induction.   Apart from CMF repression, the presence of glucose in the
growth medium results in reduced intracellular cAMP and inhibition of
inducer uptake.   These additional regulations in mixed-substrate
fermentation cause enzyme activity to approach zero as the dilution rate
is increased.   This is quite different than what was observed with
lactose alone (Ray, 1985).

   At this point it is relevant to mention the importance of the
inducer exclusion model in mixed-substrate fermentation with glucose
and a non-PTS lac inducer, like lactose.   In batch culture fermentation
it is easy to differentiate the "glucose part" from the "lactose part,"
since lactose metabolism does not start until the extracellular glucose

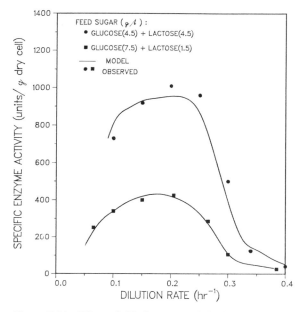

Figure 7.16   Effect of dilution rate and feed compositions on steady state enzyme specific activity
in a chemostat culture (mixed sugar medium).

level drops to a very low value.   In that case the model for the biosynthesis of lac enzymes can be used after glucose concentration is reduced below an empirically determined critical value.   In the lactose part of the diauxic growth, when the glucose level is zero, the inducer transport-exclusion model as described by eqn. [7.45] reduces to eqn. [7.17].   However, during the period when glucose concentration drops from its critical value to zero, inducer exclusion should be taken into consideration in the biosynthesis model.   But for practical purposes this period is insignificant compared to the total batch time.   In mixed-sugar chemostat culture the residual glucose concentration in the reactor is generally too low to completely shut off lactose entry, and unlike batch culture, glucose and lactose utilization cannot be readily distinguished from one another.   The transport model in this case must include glucose-mediated inducer exclusion in order to predict intracellular lactose.   Otherwise, the transport model will overpredict lactose flux into the cells.

7.19 INDUCER TRANSPORT EFFECTS

Fig. 7.17 shows the steady state experimental data and model prediction for the effect of dilution rates on intracellular lactose

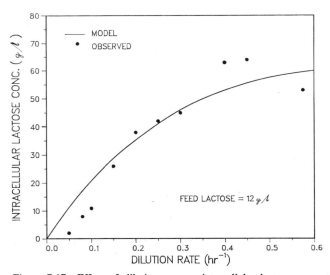

Figure 7.17   Effect of dilution rate on intracellular lactose concentration.

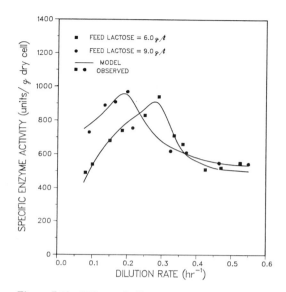

Figure 7.18 Effects of dilution rate and feed lactose concentration on steady state enzyme specific activity in chemostat cultures.

concentrations of *E. coli* growing on lactose in a chemostat culture.   The saturation trend is expected.   As the dilution rate is increased the residual substrate concentration increases in the culture and the transport model incorporates this through a balance for saturable sites.

Figure 7.18 shows the simulation curves and experimental results on steady state specific β-galactosidase activity as a function of dilution rate.   It is interesting to note that even with lactose alone the specific enzyme activity (SEA) goes through a maximum and the optimum dilution rate shifts as the feed lactose concentration is changed.   However, beyond a certain dilution rate, depending on the feed concentration, SEA becomes independent of D and, also, essentially independent of feed lactose.   When lactose is present in the culture, it is transported actively in the cells by the lac permease.   Part of the intracellular lactose is transformed to allo-lactose (glucose-1, 6-galactose), which acts as inducer, and the larger part is hydrolyzed to glucose and galactose.   Both the reactions are catalyzed by β-galactosidase.   Glucose and galactose (which is ultimately converted to glucose by the *gal* operon enzymes) are actively metabolized by the organisms.   As a result of glucose metabolism the repressive intermediary catabolite, Catabolite Modulator Factor (CMF), accumulates in the cells.

According to the model description, the accumulation of CMF is insignificant below a critical substrate utilization rate. Above the critical value, CMF exerts negative control at the level of transcription. At low dilution rates, below the critical value, only lac repressor-inducer controlled regulation at the operator site is dominating and, as the dilution rate is increased, induction predominates over cytoplasmic repression as a result of higher intracellular inducer concentration. The number of repressor molecules was assumed constant at a low level, because in a wild-type strain of *E. coli* they are synthesized constitutively and the lac i gene, which codes for the repressor protein, has a weak promoter. When the dilution rate exceeds the critical value, corresponding to the critical substrate utilization rate, the regulation of expression of the lac operon by CMF becomes the dominating factor and the enzyme activity curves drop sharply. Attainment of constant SEA level at high dilution rates is due, therefore, to the achievement of maximum intracellular CMF concentration, as predicted by the saturation kinetics of CMF production. It has been experimentally observed that the biomass yield coefficient, Y_X, is inversely proportional to the feed sugar concentration. Therefore, the dilution rate corresponding to the critical substrate utilization rate varies as the feed sugar concentration is changed.

Returning to Fig. 7.15, one sees the comparison between steady state experimental data and model predictions when both glucose and lactose are present. The parameter values are shown in Table 7.6. In these experiments, lactose to glucose ratios in the feed medium were changed, but the total sugar was kept at 6.0 g/l. As expected, a higher level of SEA was observed as the ratio was raised. As predicted by the model, the rising parts of the curves at low dilution rates are due to increased levels of induction because of higher intracellular concentration of inducer as the dilution rate is increased and the declining parts of the curves are due to domination of catabolite repression and inducer exclusion over induction.

Apart from repressive action by CMF, also observed with growth on lactose alone, as the dilution rate is increased, the presence of glucose in the external medium causes a reduced level of intracellular cAMP (indicated by eqns. [7.31] and [7.40]) and inhibition of lactose uptake (indicated by eqn. [7.47]). These additional regulations in mixed substrate fermentation, exerted by unphosphorylated enzyme IIIGlc of the PTS during glucose translocation across the plasma membrane, cause enzyme activity to approach zero as D is increased. This is quite different than what was observed with lactose alone.

Figure 7.19 clearly indicates the glucose mediated regulations of inducer entry into the cells.

7.20 TRANSIENT RESPONSE IN CHEMOSTAT CULTURES

To become better informed about the interactions of the regulatory species and the genetic control elements and to obtain indications of the strength of transient responses, the effects of transient environmental conditions on the regulation of lac operon expression were also probed in mixed sugar chemostat cultures. The organisms adjust the synthesis of macromolecules as they continue to adapt to the changing environment, and finally a new steady state is reached. The regulator and effector interact transiently with the lac control units and the control functions are expressed in specific β-galactosidase activity. Transient conditions were expressed by step changes in dilution rate and change of carbon source in the feed medium from glucose to lactose and vice versa. Detailed inspection of the responses allows the framing of a phenomenological interpretation which is completely consistent with the model that has already been described.

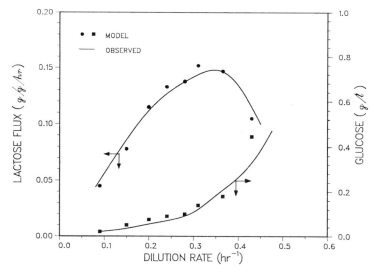

Figure 7.19   Effect of dilution rate on steady state lactose flux in a chemostat culture (mixed sugar medium).

FLOW STEP-UP

Figure 7.20 presents the transient data resulting from step-up of dilution rate from 0.095 to 0.295 hr^{-1} in a carbon-limited mixed-sugar chemostat culture.

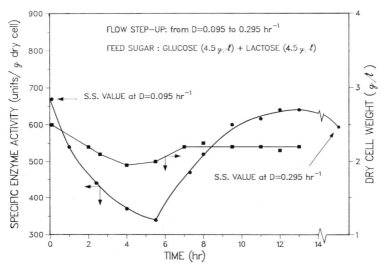

Figure 7.20   Dilution rate step-up: Enzyme specific activity in a mixed substrate chemostat culture under transient conditions.

During the transient period the specific enzyme activity underwent a cycle prior to attaining a new steady state.   The initial decline of β-galactosidase activity could be linked to combined repressive effects induced by sudden step increase in the dilution rate.   The repressive action at the onset of stepped-up conditions may be realized from the high accumulation of catabolites which cause repression of inducible enzyme synthesis.   Under carbon-limited conditions, organisms tend to derepress the synthesis of their catabolic enzymes, while simultaneously restricting the synthesis of their anabolic machinery (including ribosomes) to a level in keeping with the growth rate.   As the dilution rate was stepped up, increased amounts of sugars were taken up by the organisms and subsequently they were catabolized at a higher rate.   However, the anabolic processes did not "gear up" instantly at a similar rate, because only after a certain physiological lag would the biosynthetic machinery in the organisms respond to the changed environment. Consequently, concentration of the intracellular catabolites, including CMF, increased appreciably, causing severe inhibition of lac operon expression. At this point, the lactose transport via lactose permease (a lac operon enzyme) was reduced due to lower availability of the permease. After some time had passed in the unbalanced growth conditions, the biosynthesis rate started responding to the cellular environment and the excess catabolite pool began to deplete. Finally a balance between anabolism and catabolism was approached.

The second aspect of repression may be associated with the intracellular level of ATP, which is the substrate for the adenylate cyclase catalyzed reaction to form cAMP. It has been observed that intracellular ATP is dependent on the proton motive force across the plasma membrane (Wilson et al., 1980). Also, it has been found that the cAMP concentration inside the cells declines as a result of reduction of membrane potential (Peterkofsky and Gazdor, 1979). It is established that active lactose transport is coupled with proton co-transport, utilizing the electrochemical gradient of protons or proton motive force across the plasma membrane (Hengge and Boos, 1983). The energy of lactose transport is primarily generated by oxidation of D-lactate to pyruvate and this results in the expulsion of protons to the external medium via a membrane-bound respiratory chain.

At low dilution rate the sugar uptake rate was strictly controlled by its supply in the culture. When flow rate was stepped up, sugar uptake rate was increased. The sudden increase of lactose inflow also caused increased proton co-transport, but the proton extrusion rate could not match up with the rate of proton entry during the lag phase. This imbalance in exchange caused the proton motive force to collapse, and, as a result, intracellular cAMP dropped. At this point the low activity of lactose permease would limit lactose transport. In addition, proton extrusion would begin to "gear up" in response to the extracellular stimulus. Consequently, partially collapsed membrane potential would start to recover. (Collapse of membrane potential need not always be an unalloyed disaster, however. Papoutsakis (1986) describes the uncoupling of $\triangle$pH and metabolism in butanol/acetone fermentation where substantial redirection of the metabolic pathways toward product formation was achieved.)

Figure 7.21 presents the transient data resulting from step-up of dilution rate (from 0.1 to 0.35 hr^{-1}) in a lactose limited chemostat culture. During the transient period, when the culture was being adapted to the new dilution rate, the specific enzyme activity cycled prior to attaining a new steady state. The initial sharp decline of β-galactosidase activity could be linked to the combined effects of reduced proton motive force and excess accumulation of repressive intermediary catabolite, like CMF.   At low dilution rate (0.1 hr^{-1}) the lactose uptake rate was strictly controlled by the supply of the sugar in the culture.   When the flow rate was stepped up, an increased amount of lactose was transported into the cells, mediated by the lac permease. This sudden inrush of lactose also caused increased proton co-transport, but the proton extrusion through the respiratory chain could not match up with the proton entry during the physiological lag time. Once again, this imbalance caused the proton motive force to collapse, and consequently intracellular cAMP level dropped, resulting in increased catabolite repression.   Due to decline in β-galactosidase and presumably lac permease activity, less lactose and fewer protons would be transported into the cells.   In addition, proton extrusion through the respiratory chain would start to "gear up" in response to the stimulus arising from the changed cellular environment.   As a result, partially collapsed membrane potential would start to recover.   This is evident from the rising part of the enzyme specific activity profile.   Eventually the cAMP level would be raised and the transient situation would decay.

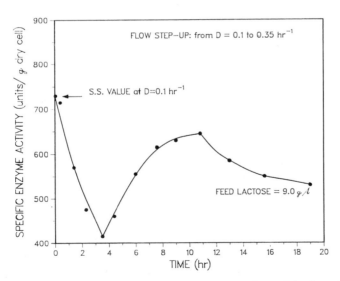

Figure 7.21   Dilution rate step-up: Enzyme specific activity in a chemostat culture under transient conditions.

Another aspect of repression at the onset of the stepped-up conditions might again be linked to the higher accumulation of repressive catabolite(s). As already mentioned, under carbon-limited conditions, organisms tend to derepress the synthesis of their catabolic enzymes, while simultaneously restricting the synthesis of their anabolic machinery (including ribosomes) to a level in keeping with the growth rate. As the dilution rate was stepped up, an increased amount of lactose was taken up by the organisms and subsequently lactose was catabolized at a higher rate. However, the anabolic processes did not "gear up" instantly at a similar rate, because only after a certain lag time, would the organisms adapt to the new environment. As a result, the concentration of the repressive catabolite(s) inside the cells increased appreciably, causing severe inhibition of lac operon expression. At this point the rate of lactose entry was reduced due to lower availability of the permease. This was indicated by the drop of biomass concentration during that period. After some time had elapsed in the unbalanced growth condition, the rate of anabolism started responding to the dynamic cellular environment and the excess intermediary catabolite pool began to deplete. Finally, balanced growth conditions were approached. Consequently, the specific activity started increasing. The β-galactosidase specific activity was found to pass through a maximum prior to reaching the new steady state (at $D = 0.35$ hr^{-1}).

RECAPITULATION

The unified model was verified with experimental results. Experiments on induction of β-galactosidase in a wild-type strain of *E. coli* B (PP01) growing in chemostat cultures on lactose (inducer substrate) as well as a mixture of glucose and lactose were used to obtain data. A NONLIN computer library program was used to estimate parameters. Molecular interactions among regulatory proteins, effectors and the genetic control sites of the DNA are increasingly revealed by research in molecular biology and genetic engineering. The model incorporates these findings and is general in the sense that it can be applied in a wide range of genetic and environmental conditions. The model was compared with experimental data from chemostat cultures and the agreement was found to be very good. None of the previous models on genetic regulations of inducible enzyme biosynthesis dealt with quantitative description of the role of CMF on catabolite repression or the role of the PTS enzyme IIIGlc in the regulation of intracellular cAMP and inducer transport. The importance of the inducer transport model has been demonstrated. Finally, the model can easily be extended to encompass the effects of genetic alterations at the regulatory sites of the lac system, cloning effects and environmental manipulations resulting in change of intracellular regulatory elements of the expression of the lac operon.

7.21 NOMENCLATURE

A, A'	lactose transport coefficients, nmol (mg protein)$^{-1}$ min^{-1}
b_1, b_2	operator index constants, g/l
B, B'	lactose transport coefficients, nmol (mg protein)$^{-1}$ min^{-1}
C, C'	lactose transport coefficients, dimensionless
D	dilution rate, hr^{-1}
D_C	dilution rate corresponding to critical substrate utilization rate
f	fraction of total specific binding sites of the lac promoter bound with RNA polymerase to form "open" complex
F	f/f_{max}, RNAP binding efficiency
$(G)_i$	intracellular glucose concentration, g/l
$(G)_o$	bulk glucose concentration, g/l
$(H^+)_{in}$, $(H^+)_o$	intracellular and bulk concentration of hydrogen ions, g/l^{-1}
K_A, K_B, K_C	dimensionless constants
K_D, K_E	constants, l/mg
K_{+E}	β-galactosidase rate constant, units/mg/hr
K_{-E}	decay constant for enzyme (β-galactosidase), hr^{-1}
K_{+M}	mRNA rate constant, mg/g
K_{-M}	decay constant for mRNA, hr^{-1}
k_m	rate constant, mg/l
k_{-m}	decay constant for catabolite modulator factor, hr^{-1}
K_{ij}, K'_{ij}	transport equilibrium coefficients
K_i, K_p	cell inhibition constants, l/g
K_1, K_2	promoter binding constants
K_{11}, K_{12}	inhibition constants, g/l
K_{S1}, K_{S2}	Monod saturation constants, g/l
$(L)_{in}$	intracellular lactose concentration, g/l^{-1}
$(L)_o$ or L_B	bulk lactose concentration, g/l^{-1}
(M)	intracellular concentration of catabolite modulator factor
p_i	concentration of inhibitory product, g/l
Q	fraction of repressor-free operator
r_{S1}	glucose uptake rate, g/l/hr
r_{S2}	lactose uptake rate, g/l/hr

r_{X1}	rate of cell mass accumulation from glucose, g/l/hr
r_{X2}	rate of cell mass accumulation from lactose, g/l/hr
Y_x	biomass yield coefficient
γ	dimensionless cAMP concentration
μ	specific cell growth rate, hr^{-1}
μ_{m1}, μ_{m2}	maximum specific cell growth rates, hr^{-1}

7.22 REFERENCES

Abou-Sabe´, M. and S. Mento, *Biochim. Biophys. Acta*, **385**, 294 (1975).

Abou-Sabe´, M., M. Burday and J. Gentsch, *Biochim. Biophys. Acta*, **385**, 281 (1975).

Abou-Sabe´, M.A., *Nature (London), New Biol.*, **243**, 182 (1973).

Alper, M.D. and B.N. Ames, *J. Bacteriol.*, **122**, 1081 (1975).

Anderson, W.B., A.B. Schneider, M. Emmer, R.L. Perlman and I. Pastan, *J. Biol. Chem.*, **246**, 5929 (1971).

Clarke, P.H., M.A. Houldsworth and M.D. Lilly, *J. Gen. Microbiol.*, **51**, 225 (1968).

Dessein, A., F. Tillier and A. Ullmann, *Mol. Gen. Genet.*,**162**, 89 (1978).

Dills, S.S., A.A. Aporson, M.R. Schmidt and M.H. Saior, *Microbiol. Rev.*, **44**, 385 (1980).

Epstein, W.L., B. Rothman-Denes and J. Hesse, *Proc. Natl. Acad. Sci. U SA*, **72**, 2300 (1975).

Gondo, S., K. Venkatasubramanian, W.R. Vieth and A. Constantinides, *Biotechnol. Bioeng.*, **20**, 1797 (1978).

Guidi-Rontani, C., A. Danchin and A. Ullmann, *Proc. Natl. Acad. Sci. USA*, **77**, 5799 (1980).

Hengge, R. and W. Boos, *Biochim. Biophys. Acta*, **737**, 443 (1983).

Imanaka, T. and S. Aiba, *Biotechnol. Bioeng.*, **19**, 757 (1977).

Iwakura, Y., K. Ito and A. Ishihama, *Mol. Gen. Genet.*, **133**, 1 (1974).

Jacob, F. and J. Monod, *J. Mol. Biol.*, **3**, 318 (1961).

Jones, T.H.D. and E.P. Kennedy, *J. Biol. Chem.*, **244**, 5981 (1969).

Joseph, E., C. Bornsley, N. Guiso and A. Ullmann, *Mol. Gen. Genet.*, **185**, 262 (1982).

Kaback, H.R., *Cell Physiol.*, **89**, 575 (1976).

Kaczorowski, G.J. and H.R. Kaback, personal communication (1979).

Kaczorowski, G.J., D.E. Robertson and H.R. Kaback, *Biochemistry*, **18**, 3697 (1979).

Kaczorowski, G.J., D.E. Robertson, M.L. Garcia, E. Padan, L. Patel, G. LeBlanc and H.R. Kaback, *Ann. N. Y. Acad. Sci.*, **358**, 307 (1980).

Kaushik, K. R., Ph.D. Thesis in Chemical and Biochemical Engineering, Rutgers U. (1981).

Kaushik, K.R., S. Gondo and K. Venkatasubramanian, *Ann. N. Y. Acad. Sci.*, **326**, 57 (1979).

Kepes, A., in "The Cellular Function of Membrane Transport," Academic Press, New York, 179 (1964).

Kepes, A., *Physiol. Veg.*, **9** (1), 11 (1971).

Kohn, L.D. and H.R. Kaback, *J. Biol. Chem.*, **248**, 7012 (1973).

Kolber, A.R. and W.D. Stein, *Nature (London)*, **209**, 691 (1966).

Kolber, A.R. and W.D. Stein, *Curr. Mod. Biol.*, **1**, 244 (1967).

Kotyk, A. and K. Janacek, "Cell Membrane Transport," Plenum, New York (1970).

Lee, J.H. and W.J. Dobrogosz, *J. Bacteriol.*, **154**, 992 (1983).

Lee, S.B. and J.E. Bailey, *Biotechnol. Bioeng.*, **26**, 1372 (1984).

Lee, S.B. and J.E. Bailey, *Biotechnol. Bioeng.*, **26**, 1383 (1984).

Lee, S.B. and J.E. Bailey, *Plasmid*, **11**, 166 (1984).

Makman, R.S. and E.W. Sutherland, *J. Biol. Chem.*, **240**, 1309 (1965).

Metzler, C.M., G.L. Elfring and A.J. McEwen, "NONLIN and Associated Programs," The Upjohn Company, Kalamazoo, Michigan (1974).

Miller, J.H., "Experiments in Molecular Genetics," Cold Spring Harbor Laboratory, New York, 367 (1972).

Mitchell, P., *Nature (London)*, **191**, 144 (1961).

Mitchell, P., *Biol. Rev. Cambridge Philos. Soc.*, **41**, 445 (1966).

Mitchell, P., *J. Bioenerg.*, **3**, 5 (1972).

Mitchell, P., *J. Bioenerg.*, **4**, 63 (1973).

Mitchell, W.J., T.P. Misco and S. Roseman, *J. Biol. Chem.*, **257**, 14553 (1982).

Monod, J., G. Cohen-Bazire and M. Cohn, *Biochim. Biophys. Acta*, **7**, 585 (1951).

Nakada, D. and B. Magasanik, *J. Mol. Biol.*, **8**, 105 (1964).

Nelson, S.O., J.K. Wright and P.W. Postma, *The EMBO J.*, **2,** 715 (1983).

Papoutsakis, T., Abstract, *Biochemical Engineering VI*, Engineering Foundation Conference, Henniker, New Hampshire (1986).

Pastan, I. and S. Adhya, *Bacteriol. Rev.*, **40**, 527 (1976).

Perlman, R. and I. Pastan, *J. Biol. Chem.*, **243**, 5420 (1968).

Peterkofsky, A. and C. Gazdor, *Proc. Nat. Acad. Sci. USA*, **76**, 1099 (1979).

Peterkofsky, A. and C. Gazdor, *Proc. Nat. Acad. Sci. USA*, **71**, 2324 (1974).

Pichon, J.L., C. Coeroli and G. Marchis-Mouren, *Mol. Gen. Genet.*, **150**, 257 (1977).

Pirt, S.J., "Principles of Microbe and Cell Cultivation," John Wiley and Sons, New York (1975).

Postma, P.W. and S. Roseman, *Biochim. Biophys. Acta,* **457**, 213 (1976).

Postma, P.W., A. Schuitema and C. Kwa, *Mol. Gen. Genet.*, **181**, 448 (1981).

Ray, N.G., W.R. Vieth and K. Venkatasubramanian, *Ann. N.Y.Acad. Sci.*, **469**, 212 (1986).

Ray, N.G., Ph.D. Thesis in Chemical and Biochemical Engineering, Rutgers U. (1985).

Saier, M.H., *Bacteriol. Rev.*, **41**, 856 (1977).

Saier, M.H., M.J. Novotny, D. Comeau-Fuhrman, T. Osumi and J.D. Desai, *J. Bacteriol.*, **155**, 1351 (1983).

Schaffer, E.J. and C.L. Cooney, *Appl. Environ. Microbiol.*, **43**, 75 (1982).

Scholte, B.J., A.R. Schuitema and P.W. Postma, *J. Bacteriol.,* **148**, 257 (1981).

Sherwood, T.K., R.L. Pigford and C.R. Wilke, "Mass Transfer", McGraw Hill (1975).

Spassky, A., S. Busby and H. Buc, *The EMBO J.*, **3**, 43 (1984)

Strauss, H.S., R.R. Burgess and T.R. Reurd, Jr., *Biochemistry*, **19**, 3504 (1980).

Stroobant, P. and H.R. Kaback, *Proc. Nat. Acad. Sci. USA*, **72**, 3970 (1975).

Takahashi, M., B. Blazy and A. Baudras, *Biochemistry*, **19**, 5124 (1980).

Takahashi, M., B. Blazy, A. Baudras and W. Hillen, *J. Mol. Biol.*, **167**, 895 (1983).

Toda, K., *Biotechnol. Bioeng.*, **18**, 1117 (1976).

Ullmann, A. and A. Danchin, "Advances in Cyclic Nucleotide Research," P. Greengard and G.A. Robison (Eds.), **15**, 1 (1983).

Ullman, A., F. Tillier and J. Monod, *Proc. Natl. Acad. Sci. USA*, **73**, 3476 (1976).

Venkatasubramanian, K., A. Constantinides and W.R. Vieth, Fifth International Fermentation Symp. Abst., Springer-Verlag, Berlin, 97 (1976).

Vieth, W.R., K. Kaushik and K. Venkatasubramanian, "Enzyme Engineering," Plenum Press: New York, **6**, 45 (1982).

Vieth, W.R., K. Kaushik and K. Venkatasubramanian, *Biotechnol. & Bioeng.*, **24**, 1455 (1982).

Wilson, D.M., M. Kusch, J.L. Flag-Newton and T.H. Wilson, *FEBS Letters*, **117** (suppl), K37 (1980).

Wright, L.F., D.P. Milne and C.J. Knowles, *Biochim. Biophys. Acta.*, **583**, 73 (1979).

Yagil, G. and E. Yagil, *Biophys. J.*, **11**, 11 (1971).

Zubay, G., *Cell Biol.*, **3**, 154 (1980).

We were thirty miles from
        Coldspring,
forget it I never shall!
What a churning tide we rode one
        day
to reach our home canal.

Song for the Cape May "Mosquito Fleet,"
Adapted from Trad. Ballad, "The Erie
Canal"

8

INDUCIBLE RECOMBINANT CELL CULTURES AND BIOREACTORS

8.0 INTRODUCTION

In several ways, this chapter represents a fusion of the ideas taken up in Chapters 5 and 7. The structural model of the lac operon presented in Chapter 7 illustrates control at the gene level; a natural challenge to the biochemical engineer is to extend or develop such models to encompass the variations produced by molecular cloning (e.g., Shuler; Ryu; Ray; Dhurjati; 1986). While we will continue to focus on gene expression in the form of the enzyme, β-galactosidase, it is important to note that the lac promoter is widely used as a genetic regulatory unit for initiation of transcription for cloned genes in a variety of multicopy recombinant plasmids, so the modeling effort to be presented has a certain generality.

In the work described here, the host was $E.\ coli$ MBM7061 (z^-, y^+), harboring the plasmid pMLB1108 which is lac $i^q z^+$. Thus, while the host strain does contain the gene which codes for the permease (y^+), it is deficient in the z gene for expression of β-galactosidase, in the normal position on the chromosome. This deficiency is made up by the plasmid pMLB1108 (see Fig. 8.1), which contains the proximal half of the lac operon; i.e., the repressor gene (i^q) and the gene coding for β-galactosidase (z^+). The q mutation causes the repressor protein to be overproduced, allowing better control for the switching on and off of protein synthesis in chemostats and in immobilized cells. Thus, one has a very flexible genetic system to study strategies for maintaining plasmids in genetically engineered systems.

The second rather important aspect of this work is the comparative study of plasmid-bearing cells in free suspension cultures and in immobilized states. So-called "plasmid-shedding" occurs as a consequence of natural selection; i.e., competitive growth of wild-type and recombinant strains on the available substrate. If cell division is blocked (by immobilization), differential reproduction cannot take place. Thus, immobilized recombinant cell cultures are expected to be more stable (Dykhuizen and Hartl, 1983; De Taxis du Poet et al., 1986).

However, weighing against this advantage is the possibility that immobilized cells may not overproduce the desired metabolite as actively. But that, too, is mitigated by the flexible metabolic switching policies afforded by the inducer/repressor combination which, at least in theory, could be applied at an optimal time in the growing cycle.

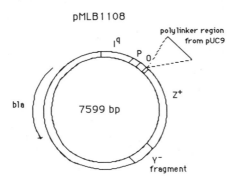

Figure 8.1 Genetic Map of pMLB1108.

8.1 STRAIN DEVELOPMENT

The initial approach was to clone the entire lac structural gene into a multicopy expression vector, so that the analysis of enzyme biosynthesis in wild-type *E. coli* could be readily applied to the recombinant system as well.

In keeping with this objective, a plasmid was constructed possessing the following characteristics:

- contains entire structural region of the lac operon (*z, y* and *a*) coding for production of β-galactosidase, lac permease, and lac transacetylase, respectively.
- contains the lac promotor-operator region.
- contains a genetic marker for antibiotic resistance.
- plasmid possesses a "relaxed" genetic control mechanism (*ColEl* origin), making plasmid replication continuous (Scott, 1984).

The lac structural gene (*z, y* and *a*) was removed from the plasmid pMC81 (Casadaban and Cohen, 1980) (see diagram below) by performing a double restriction enzyme digestion with *HindIII* and *BamHI* (Maniatis et al., 1982). The multicopy vector pUC9 (Vieira and Messing, 1982) (also shown below) was similarly treated to open the polylinker region, producing "sticky" ends with the same orientation as the structural gene DNA segment. Both portions of DNA were isolated using low melting point agarose gel electrophoresis. The appropriate size bands were excised from the gel, the agarose melted at 70°C and combined with ligation buffer and T4 DNA ligase to seal the DNA fragments together, thereby producing the desired plasmid. The cloning procedure is illustrated in Figure 8.2.

The newly synthesized plasmid (pKB1) was used to transform two different *E. coli* strains, JM103 and TB1. The first of the two host strains, JM103 (Messing et al., 1981), is a *laci*^{q} strain, meaning that it overproduces the lac repressor protein which in turn blocks transcription of the lac structural gene. A *laci*^{q} host can produce as much as ten times the amount of repressor present in a normal cell (Bailey et al., 1983). Overproduction of the lac repressor by the resulting recombinant cells is necessary to provide inducible enzyme biosynthesis. The second host (TB1) is a Restriction (-) derivative of JM83. It was obtained commercially from Bethesda Research Laboratories. The TB1 host should allow constitutive enzyme biosynthesis to take place in the resulting recombinant cells, because the lac repressor gene is absent. However, unlike the recombinant cells obtained using the JM103 host, no means is provided for controlling the rate of gene expression through an induction.

Both strains contain Δ (*lac pro*) deletions; i.e., neither host strain can metabolize lactose or synthesize proline. This gives the host cells a Lac (-) phenotype, meaning that without plasmid they are incapable of producing active β-galactosidase. Therefore, lactose-containing medium selects for recombinant cells.

The transformed cultures were spread on McConkey-Ampicillin plates, then incubated overnight at 37°C. The recombinant colonies were red, although many white colonies were present. Red colonies were restreaked onto fresh plates for further purification. Restreaking of the KB2 and KB3 recombinant strains was repeated several times until no further segregation of red and white colonies resulted.

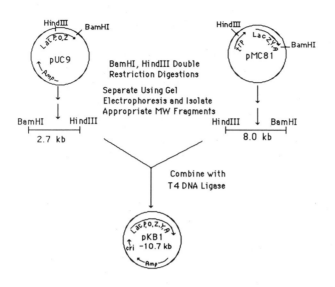

Figure 8.2   Cloning Procedure.

The plasmid DNA of the KB2 and KB3 strains was isolated, digested with the restriction enzymes *HindIII* and *BamHI* and the fragments separated by agarose gel electrophoresis. The fragmentation pattern showed the presence of the 8 kb insert corresponding to the lac structural gene.

Both the KB2 and KB3 recombinant strains were evaluated for β-galactosidase production in small-scale culture. The two strains produced the enzyme, indicating that the plasmid contained the *lac z* gene cloned into it; however, the expression levels of enzyme were substantially lower than expected. Similar problems have been noted by other researchers after cloning the lac permease gene into plasmids (Berman and Jackson, 1984).

In the development of a recombinant strain to be used in subsequent bioreactor studies, it was necessary to have a host possessing the lac permease gene so that lactose could be metabolized. The host therefore had to be *lac z⁻, y⁺* so that the *lac z⁺* plasmid and host could be easily distinguished when grown on indicator plates. A plasmid/host system was chosen which meets these requirements, and provides for inducible biosynthesis of β-galactosidase. The plasmid, pMLB1108 (shown in Figure 8.1), was developed by Michael L. Berman at the National Cancer Institute. It contains *lac iq o, p and z* genes. The plasmid repressor gene overproduces the lac repressor protein, thereby providing the ability to regulate protein biosynthesis by controlling the inducer (i.e., lactose or IPTG) concentration.

The host is a *lac z⁻, y⁺* strain, MBM7061 (Berman and Jackson, 1984). The permease activity was obtained through lysogenic incorporation of the transducing phage λp1048 in the *E. coli* host strain MC4100 (Casadaban, 1976). The resulting recombinant strain was tested in small-scale culture and found to possess the desired characteristics. It was used for all subsequent bioreactor studies. Therefore, all experimental results were obtained using this recombinant strain.

8.2 FRACTION OF PLASMID CONTAINING CELLS

The occurrence of plasmid segregational instability in free and immobilized cell bioreactor systems was monitored by evaluating the phenotypical characteristics of the cells in the effluent stream. Two plating techniques were used for this purpose. Both procedures involved performing serial dilutions of the reactor effluent using sterile isotonic saline to obtain a final cell concentration of 100-300 cells per 100 μl which was the volume of diluted sample injected onto the surface of the petri dish. The cells were evenly spread over the surface of the plate using a sterile bent glass rod. The exact dilution used for a given sample varied, depending upon the cell concentration in the effluent, but typically involved three successive 1:100 or 1:50 dilutions. The plates were spread in duplicate and sometimes triplicate for greater statistical accuracy. The plates were then incubated overnight at 37°C and the colonies counted.

The first plate medium used was LB with and without ampicillin. The nonselective LB plates gave the total cell count in the reactor effluent, while the ampicillin (50 μg/ml) containing plates gave the population of ampicillin resistant cells in the effluent, which ideally should correspond to the number of plasmid containing cells. The fraction of ampicillin resistant cells was obtained by dividing the ampicillin resistant cell count by the total cell count.

The second plate medium used was McConkey agar plates without any ampicillin. The recombinant cells possess a Lac⁺ phenotype, giving rise to red colonies on McConkey plates, while the Lac⁻ wild-type cells produce white colonies. The recombinant cell fraction was therefore obtained from the ratio of red colonies to the total number of colonies on the McConkey agar plates. The ampicillin resistance of the red (recombinant) colonies was periodically checked by replica plating both red and white colonies on McConkey plates containing 50 μg/ml ampicillin to assure that the red colonies were also ampicillin resistant while the white colonies did not grow at all on ampicillin containing plates. The McConkey agar plates had two distinct advantages over the LB plates.

1. There is no ambiguity in discerning between wild-type and recombinant cells on the same plate. One cannot conclusively tell which cells are wild-type on LB plates without replating colonies.

2. Both recombinant and wild-type cell populations are counted from the same plate, giving much greater accuracy in calculating the recombinant cell fraction. This is especially important at the beginning of a plasmid segregational instability study. From a statistical standpoint, the error between sets (plates) is eliminated when using McConkey plates. The error between plates is largely the result of unequal sample volumes, and unevenness of cell sample spreading.

8.3 PLASMID COPY NUMBER DETERMINATION

Plasmid copy number was estimated for both free and immobilized cell cultures by determining the quantity of Plasmid DNA obtained from a known number of cells. For free cell cultures, a single colony was used to inoculate an overnight culture in LB/ampicillin (100 μg/ml) medium. 0.5 ml of culture was pelleted and plasmid DNA was extracted as described below. Dilutions of the remaining culture were plated on LB/ampicillin (50μg/ml) and the cell number determined from counting colonies. Frozen samples of immobilized cells were used by resuspending the samples in 2 mls of LB; 200 μl aliquots were taken and the plasmid isolated as described below. Serial dilutions of the resuspended samples were made and cells were plated on LB ampicillin plates to determine the cell number in the sample.

Plasmid DNA was obtained by the alkaline lysis method as described by Maniatis et al. (1982). Cells were pelleted for one minute in an Eppendorf centrifuge and the supernatant removed by aspiration. The pellet was resuspended in 100 μl 50 mM glucose, 10 mM EDTA, 25 mM Tris-HCl (pH 8.0) with 4 mg/ml lysozyme added and incubated at room temperature for 5 minutes. 200 μl freshly prepared 0.2 N NaOH and 1% SDS were added and the tubes were mixed by inverting. Following 5 minute incubation on ice, 150 μl ice cold potassium acetate solution (3 M with respect to potassium, 5 M with respect to acetate, pH

4.8) was added. The samples were mixed by gently vortexing the inverted tube for 10 seconds and then incubated on ice for 5 minutes. The tubes were centrifuged for 5 minutes in an Eppendorf centrifuge at 4°C. The supernatants were decanted to fresh tubes. 400 μl phenol/chloroform (1:1) was added to each tube and the contents mixed by vortexing. After 2 minutes of centrifugation in an Eppendorf centrifuge, the supernatant was transferred to a clean tube. 800 μl ethanol was added and incubated at room temperature for 2 minutes. After 5 minutes of centrifugation, the supernatant was removed by aspiration. The pellet was washed with 1 ml of 70% ethanol, recentrifuged and the supernatant removed by aspiration. The pellets were dried in a speed vac and then dissolved in 50 μl 10 mM Tris-HCl, 1 mM EDTA (pH 8.0) containing DNase-free pancreatic RNase (20 μg/ml) and incubated at 37°C for 30 minutes.

A 5 μl sample of the extracted plasmid was digested with 1 μl Bam HI (Bethesda Research Laboratories, Gaithersberg, Maryland)   in buffer supplied by the manufacturer for 1 hour at 37°C. Dilutions of the linearized plasmid were electrophoresed in 1% agarose gel in TBE buffer (Maniatis et al., 1982). Plasmid concentration was estimated by comparison to standard plasmids of known concentration. The plasmid copy number was calculated from the plasmid DNA concentration and the corresponding number of cells used in the DNA isolation procedure. The two free cell samples gave plasmid copy numbers of 9.2 and 12.9 plasmids/cell, while the five immobilized cell samples gave plasmid copy numbers of 11.7, 8.2, 9.0, 14.1 and 10.5 plasmids/cell. The average of the free cell samples was 11.1 plasmids/cell, while the average copy number of the immobilized cell samples was 10.7 plasmids/cell.

8.4 IMMOBILIZED CELL CONCENTRATION

The recombinant cell culture (pMLB1108 x MBM7061) was immobilized in 1.1% alginate (with no added silica) according to a standard procedure. Three flasks, each containing 35 ml of immobilized phase and 80 mls of 5 g/l lactose tris-buffered minimal medium were incubated at 37°C, 200 rpm for 48 hours. The liquid broth was exchanged three times for fresh medium during this incubation period. The liquid broth was decanted and approximately 75 mls of sugar-free phosphate-buffered minimal media were added to each flask. The flasks were shaken at 200 rpm for one hour at room temperature to dissolve the gel beads, freeing the immobilized cells. The volume of liquid in each flask was measured in sterile 100 ml graduated cylinders. Serial dilutions of the samples were made (3 x 1:100), then 100 μl samples were spread in triplicate on LB plates. The plates were incubated overnight at 37°C, and the number of colonies on each counted. The cell counts were averaged and multiplied by the appropriate dilution factors to obtain the immobilized cell concentration in each flask. The cell concentrations in the three flasks were 2.92 g/l, 3.17 g/l and 3.66 g/l. The average of these three values (3.25 g/l) was used for all calculations involving immobilized cell concentration. The remaining cell samples in the graduated cylinders were transferred to sterile 35 ml centrifuge tubes and pelleted at 5000 rpm using an SS-34 rotor. These pelleted cell samples were frozen at -20°C and used in the plasmid copy number determination procedure described previously.

8.5 CONTINUOUS FREE CELL BIOREACTOR STUDIES

Continuous free cell bioreactor (CSTR) performance was studied by obtaining steady state cell, substrate and product concentration data as a function of dilution rate. All steady state data were obtained using a Multigen fermentor (New Brunswick Scientific) with a 1.0 liter working volume. (Other details are given by K. Bailey, 1986). The carbon and inducer sources were varied to observe differences in cell growth and gene expression kinetics.

The results from the CSTR bioreactor studies, using lactose (5 g/l) as the carbon source and inducer, are shown graphically in Figures 8.3 and 8.4. Cell washout from the bioreactor occurred at a dilution rate of 0.97 hr^{-1}. The corresponding substrate (lactose) profile shows a cell/substrate yield factor of 0.3 g dry cell weight/g lactose consumed. The maximum β-galactosidase concentration was 5000 units/g dry cells, nearly an order magnitude greater gene expression than with wild-type *E. coli* (Chapter 7), which occurred at a dilution rate of 0.31. The enzyme productivity displayed a maximum at 2950 units/l/hr, occurring at a dilution rate of 0.8 hr^{-1}. Near the cell washout condition, the productivity value rapidly goes to zero due to the loss of cells. The specific productivity data, shown in Figure 8.4, provide an indication of the rate of β-galactosidase biosynthesis inside an individual cell. This rate increases almost linearly with cell growth rate. This type of relationship is expected for production of any protein which is necessary for cell growth.

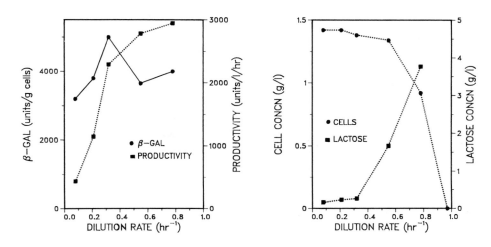

Figure 8.3   Steady state CSTR bioreactor performance: lactose feed.

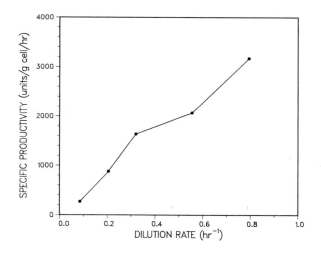

Figure 8.4   Steady state CSTR bioreactor performance: lactose feed, specific productivity data.

The steady state CSTR bioreactor results using minimal glucose media (5 g/l) are shown in Figure 8.5.   Cell washout, in this case, occurs at a dilution rate of 1.05 hr^{-1}, slightly higher than that observed for lactose media, indicating that the recombinant cells have a higher maximum specific growth rate on glucose media.   The level of gene expression is extremely low at all dilution rates;   the maximum β-galactosidase concentration was 43 units/g cells, which occurred at a dilution rate of 0.52 hr^{-1}.   This represents less than 1% of the enzyme activity observed when grown on lactose minimal media.   The foregoing results establish both the inducibility and the susceptibility to catabolite repression of the recombinant cell cultures.

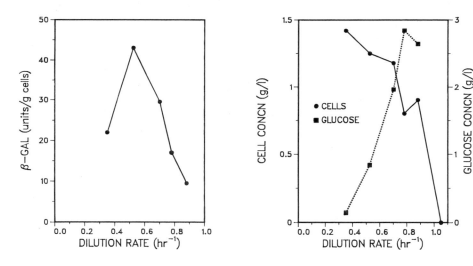

Figure 8.5   Steady state CSTR bioreactor performance: glucose feed.

8.6 IMRC BIOREACTOR

An IMRC bioreactor is, usually, a three-phase packed or fluidized bed reaction system, possessing solid, liquid and gas phases. An IMRC bioreactor may be operated in either mode.   The disadvantage in using a packed bed reactor is that channeling of air or medium through the bed can   occur, leading to stagnant zones in the reactor with very poor mass transfer rates.   Most often gas (air) bubbles are used to provide mixing of the solid and liquid phases. However, this type of three-phase system is difficult to scale up which makes it difficult to obtain consistent results among reactor runs, due to differences in fluidization characteristics. Aeration of the fermentation broth in a separate vessel and rapid recycle of the oxygenated liquid to the IMC reactor column provides more gentle and predictable fluidization while obtaining excellent mass transfer (Karkare et al., 1985). As a result, scale-up of the IMC reactor becomes much more predictable.

The bioreactor design (K. Bailey, 1986) actually used for immobilized cell reactor studies is shown schematically in Figure 8.6. The reactor is a three column system, composed of the IMRC reactor column (shown right), the liquid phase oxygenation column (shown left), and the gas-liquid disengagement column (located in the center). The main feature of this reactor system is that the liquid phase is oxygenated in a separate vessel from the column containing the gel particles. The advantages of this design are:

- The rates of fluidization and oxygenation can be controlled independently.
- The mixing pattern inside the IMRC reactor column is more thorough, yet more gentle due to lower shear forces, thereby reducing the rate of bead attrition.

The total system volume (liquid + gel particles) varied between 650 and 750 mls, depending on the liquid level in the IMRC reactor. The

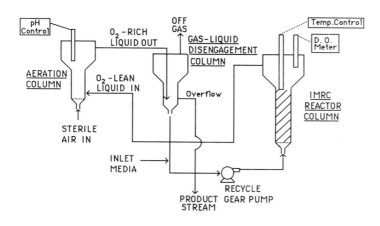

Figure 8.6 IMRC reactor design.

temperature was maintained at 37°C.   Approximately 200 mls of gel beads were used in each reactor run.   The cell broth was recycled between the IMRC reactor column and the oxygenation unit; a liquid recycle rate of approximately 700 ml/min was maintained.

When using a phosphate buffered medium, it was unnecessary to control the pH in the reactor system.   The phosphate salts had an ample buffering capacity.   The *E. coli* culture has a pH of approximately 6.80 when grown to stationary phase in a phosphate buffered medium.   In contrast, when grown in tris buffered medium, a stationary phase *E. coli* culture has a pH of 4.20.   Tris buffered medium had to be used with calcium alginate gel, because high phosphate concentrations dissolve the alginate material, by chelating the calcium ions.   It was therefore necessary to control the system pH when using tris buffered media.

8.7 IMRC BIOREACTOR STUDIES

The IMRC bioreactor system was operated using tris-buffered minimal lactose media as the inlet feed, and calcium alginate as the gel

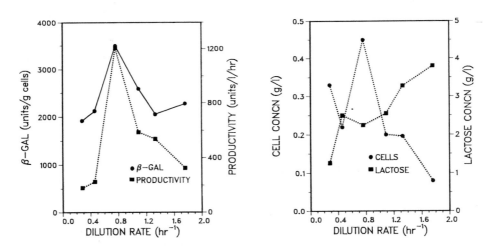

Figure 8.7   Steady state IMRC bioreactor performance: lactose feed.

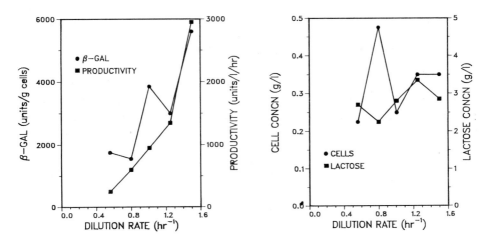

Figure 8.8   Steady state IMRC bioreactor performance: lactose feed.

matrix material.   Steady state effluent concentrations of cell, lactose
and β-galactosidase were obtained as functions of dilution rate, which is
based on the total (liquid + gel phase) volume of the three column
bioreactor system.   A protracted experiment was performed twice, at
one week intervals, using the same bioreactor system.   Experimental
results are shown in Figures 8.7 and 8.8, respectively.

These results show that the IMRC bioreactor system can be
operated at a dilution rate as high as 1.8 hr^{-1} without incurring the cell
washout problem observed in the free cell system at a dilution rate of
0.97 hr^{-1}.   In the second trial, higher values for maximum β-
galactosidase concentration (5600 units/g cells) and reactor
productivity (2900 units l^{-1} hr^{-1}) were obtained than in the first one.
This revealed the presence of a substrate limitation to cell growth other
than carbon source (lactose) in the former.   The tris-buffered growth
medium has a phosphate concentration of 1.0 mM;   however, calcium
chloride has to be added to the growth medium to prevent the gel
particle beads from swelling and dissolving.   But phosphate anion
chelates calcium ions;   the net result is that limited free phosphate is
available to the suspended cells and still less is available to the
immobilized cells, due to mass transfer limitations.

Nonetheless, in spite of the lack of smoothness in the IMRC bioreactor profiles, the IMRC bioreactor system did demonstrate that product concentrations and volumetric productivities similar to the free cell CSTR bioreactor system were obtainable;   however, this might be vastly improved by increased cell density.   The specific productivity data, shown in Figure 8.9, indicate that the rate of enzyme biosynthesis in the immobilized cells is approximately equal to the corresponding rate of enzyme production in the CSTR bioreactor study (Figure 8.4) at similar dilution rates;   however, at dilution rates above the maximum specific growth rate there is a two to three-fold improvement for the IMRC bioreactor productivity over the CSTR bioreactor case.

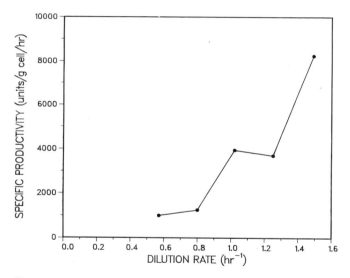

Figure 8.9   Steady state IMRC bioreactor performance: lactose feed, specific productivity data.

8.8 RECAPITULATION OF THE GENE EXPRESSION SYSTEM

Regulation of protein biosynthesis is crucial for a growing cell to regulate the concentration of catabolic and anabolic enzymes and ensure efficient use of available resources.   Protein production involves two discrete steps:

1. Transcription of the structural gene coding for the protein to produce messenger RNA (mRNA).

2. Translation of mRNA into protein by ribosomal enzymes.

In most cases, control of protein biosynthesis occurs at the first step, by controlling the frequency of transcription initiation. Initiation may be controlled by the promoter gene which binds RNA polymerase, or by the operator gene, which, if bound by repressor protein, blocks transcription from the promoter site. A multitude of different promoter-operator genes exists, so the exact control mechanism is dependent upon which control region governs expression of the structural gene.   The lac operon system is perhaps the best understood and was the system chosen in the studies conducted in our laboratory, as previously described in Chapter 7.

The lac operon is shown schematically in Figure 8.10. Control of mRNA synthesis for β-galactosidase takes place at two independent sites.

1. The repressor gene (i) constitutively produces a protein (R) which, in the absence of inducer (lactose), binds to the operator gene (o), blocking transcription from the promoter.

2. In the presence of high levels of cyclic adenosine monophosphate (cAMP), a complex forms between cAMP and the catabolite receptor protein (CRP) which then binds to the P_I region of the promotor gene, facilitating RNA polymerase binding at the P_{II} site (de Crombrugghe et al., 1984).

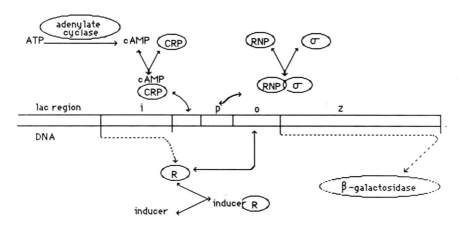

Figure 8.10   Schematic of lac operon transcription control system.

Thus negative regulation takes place at the operator gene while positive regulation occurs at the promoter P_I site. A high lactose concentration is sufficient to prevent the repressor protein from binding to the operator gene, but in the presence of glucose, low cAMP levels prevent the CRP-cAMP complex from forming and binding to the promoter. As a result, extremely low levels of β-galactosidase are synthesized by the cell. There is evidence that the cAMP-CRP complex is not the sole mediator of catabolite repression (Wanner et al., 1978). This is confirmed by the fact that catabolite repression can take place in *crp* mutants, in the absence of a functional cAMP-CRP complex (Guidi-Rontani et al., 1980). An intermediate in the metabolism of glucose is thought to bind free RNA polymerase (RNAP), inhibiting the binding of RNAP to the P_{II} site of the promoter gene (Dessein et al., 1978). This intermediate has been given the name "catabolite modulating factor."

As described in Chapter 7, the effects of catabolite modulating factor have been incorporated into a rigorous mathematical model of β-galactosidase biosynthesis in wild-type *Escherichia coli* (Ray et al., 1987). The model provides a quantitative description of active inducer transport (Kaushik et al., 1985) and the role of PTS (phosphoenolpyruvate dependent sugar transport system) enzymes in the regulation of cAMP levels and in inducer exclusion.

Genetic engineering technique allows researchers to modify promotor and operator functions and to use them to govern expression of a foreign gene by cloning it into the structural gene region of the operon. The lac operon is used frequently in many plasmid constructions, either in its entirety or as a portion. Since its expression is so well characterized, it is convenient to use the lac operator-promoter region to regulate expression of cloned genes. The pUC plasmids have been developed for this purpose (Vieira and Messing, 1982).

A structured model for lac promoter-operator function in wild-type and high copy number plasmids has been developed (Lee and Bailey, 1984a, b). The effect of plasmid copy number on protein production was modeled with results showing that the rate of transcription with respect to plasmid copy number deviates from linearity at high copy number, as one would expect.

8.9 BIOREACTOR MODEL

The mathematical model is, once again, written in terms of its component parts. The important variables, characterizing the state of the system, are the recombinant and plasmid-free cell concentrations in the bulk phase, bulk phase substrate concentration and the concentration of the enzyme β-galactosidase in the reactor effluent. As the immobilized phase makes a substantial contribution to the bulk phase values of these variables, it is necessary to evaluate the rates of cell division and substrate consumption in the immobilized phase as well. Assuming that there is no net accumulation of substrate (lactose) in the immobilized phase, the diffusion of substrate through the spherical gel particle must equal the rate at which it is consumed. This is expressed by equation [8.1].

Intraparticle substrate diffusion-reaction balance equations

$$D_{eff} \frac{d}{dr} [r^2 \, dS / dr] = r [S] \qquad [8.1]$$

This differential equation may be solved by using two of the following boundary conditions (the first being essential):
Surface Condition:

$$D_{eff} \frac{dS}{dr} = k_s [S_F - S], \quad r = R \qquad [8.2]$$

Interior Condition:

$$\frac{dS}{dr} = 0, \quad r = r^* \qquad [8.3]$$

or:
$$S = 0 = \frac{dS}{dr}, \quad r = r^* \qquad [8.4]$$

Equations [8.1] to [8.4] are useful when the carbon source within the gel particle is the limiting nutrient, controlling the rate of immobilized cell growth and division. To allow for the possibility that another nutrient limits cell growth (e.g., oxygen or phosphate) or in the absence of sufficiently detailed information regarding (carbon) substrate diffusion to the immobilized cells, it is advantageous to express the immobilized cell growth rate as a fraction of the free cell growth rate (e.g., as in eqn. [5.68]). The reaction term, r(S) may be evaluated using a simple substrate utilization model to relate it to the rate of cell production.

$$r\,(S)\; = \;\frac{1}{Y_x}\,r\,(X) \qquad\qquad [8.5]$$

where r(X) is obtained from a modified Monod model.

$$r\,[X]\; = \;\frac{\mu_{max}\cdot S\cdot X}{S + K_S\,[1 + K_i\,X]} \qquad\qquad [8.6]$$

IMRC reactor balance equations

Recombinant and plasmid-free cell concentrations may be obtained from *component* balances for each species in the following reactor model. The cell concentration in the immobilized phase is assumed to remain constant with respect to time and dilution rate. The idea here is that live cells grow to saturate a pellicular layer near the particle surface, accessible to substrate. Any additional replication is then released to the bulk. The liquid (free) phase concentrations of recombinant and wild-type cells may be calculated for a given set of operating conditions. The net rate of free-cell production of each species is the sum of contributions from the liquid and immobilized phases, minus the rate at which cell removal occurs from the bioreactor. The following list of assumptions is made in the development of the subsequent *component* balance equations for recombinant and plasmid-free cells.

1. All cells reproduce by binary fission.
2. All recombinant cells contain the same number of plasmids corresponding to the mean plasmid copy number.
3. Plasmid is divided randomly between daughter cells at the time of cell division.
4. No plasmid-free cell or its progeny can regain the plasmid.

Recombinant   cell   balance:

$$\frac{dX_F^+}{dt} = [1 - \Theta_I]\,\mu_I^+\,X_I\,\phi_I\,\frac{1 - \epsilon}{\epsilon} + [1 - \Theta_F]\,\mu_F^+\,X_F^+ - \frac{D}{\epsilon}\,X_F^+ \qquad [8.7]$$

The first term on the right hand side of eqn. [8.7] represents release of plasmid-bearing cells from the immobilized phase to the bulk phase; the other two terms are the customary ones for chemostats, expressing plasmid-bearing cell growth and removal in the reactor effluent.

Plasmid-free   cell   balance:

$$\frac{dX_F^-}{dt} = [\Theta_I\,\mu_I^+\,X_I\,\phi_I + \mu_I^-\,X_I\,(1 - \phi_I)]\,\frac{1 - \epsilon}{\epsilon} + \Theta_F\,\mu_F^+X_F^+ + \mu_F^-\,X_F^- - \frac{D}{\epsilon}\,X_F^- \qquad [8.8]$$

The composite term appearing first on the right hand side of eqn. [8.8] represents release of plasmid-free cells, while the second term represents cell reversion in the free suspension culture.

Substrate   balance:

$$[\epsilon + \xi_I\,(1 - \epsilon)]\,\frac{dS_F}{dt} = D\,[S_o - S_F] - \frac{1}{Y_X}\,r[X] \qquad [8.9]$$

where:

$$r[X] = [\phi_I\,\mu_I^+ + [1 - \phi_I]\,\mu_I^-]\,X_I[1 - \epsilon] + \mu_F^+X_F^+\epsilon + \mu_F^-\,X_F^-\,\epsilon \qquad [8.10]$$

The immobilized cell growth rate of each species can be related to the corresponding free cell growth rate as follows:

$$\mu_I^+ = \eta_I \cdot \mu_F^+ \quad \text{and} \quad \mu_I^- = \eta_I \cdot \mu_F^- \qquad [8.11]$$

The terms η_I and ξ_I are bioreaction and substrate distribution effectiveness factors, respectively. These effectiveness factors may be evaluated using the following equations:

$$\eta_I = \frac{\int_{r*}^{R} r[S] \cdot 4\pi r^2\,dr}{[\int_{r*}^{R} 4\pi r^2\,dr] \cdot r[S_F]} \qquad [8.12]$$

and

$$\xi_I = \frac{\int_{r*}^{R} S \cdot 4\pi r^2 \, dr}{[\int_{r*}^{R} 4\pi r^2 \, dr] \cdot S_F}$$ [8.13]

Equations [8.12] and [8.13] are included here for the sake of completeness. In the evaluation of steady state bioreactor kinetics, the term containing ξ_I drops out of the analysis. In the IMRC bioreactor simulations, shown later, all of the experiments were essentially steady state with respect to substrate concentration, and η_I was treated as a parameter and evaluated by simulation to match the predicted cell concentration to that observed experimentally.

The fraction of plasmid-containing cells in the immobilized phase, ϕ_I, may be evaluated using the following relation developed by Bailey et al. (1983). The fraction of plasmid-containing cells in the immobilized phase is given by:

$$\frac{1}{\phi_I[t]} = 1 + \frac{\theta[1 - \omega]}{[1 - \theta][2 - \omega]} [1 - [2^\alpha / 2 - \theta]^m]$$ [8.14]

where

$$\omega = [2 - \theta]^{1/\alpha}, \quad \theta = 2^{1 - N_p}, \quad \alpha = \mu_I^- / \mu_I^+, \quad m = t / G_I^+$$

The recombinant cell fraction in the bioreactor effluent is the same as the fraction of recombinant cells in the liquid phase. It may be calculated directly from the integrated values of X_F^+ and X_F^-.

The fraction of plasmid-containing cells in the liquid phase is given by:

$$\phi_F = \frac{X_F^+}{X_F^+ + X_F^-}$$ [8.15]

GENE EXPRESSION KINETICS

Once the recombinant cell concentration in the bulk phase is known, lac operon kinetics are used to evaluate the concentration of β-galactosidase in the effluent stream (Ray et al., 1986). The

concentration of β-galactosidase is dependent upon the concentration of mRNA which codes for the enzyme. The β-galactosidase mRNA concentration is dependent upon the rate of initiation of transcription.
β-galactosidase mRNA balance:

$$\frac{d\,[mRNA \cdot X_F^+]}{dt} = K_{+M} \cdot N_p \cdot F \cdot Q \cdot \mu_F^+ \cdot X_F^+ - [K_{-M}+D/\epsilon]\,[mRNA \cdot X_F^+] \quad [8.16]$$

where the *gene copy number*, N_p, appears explicitly. The concentration of β-galactosidase can readily be calculated from the mRNA concentration, based on the translation kinetics.
β-galactosidase enzyme balance:

$$\frac{d[E \cdot X_F^+]}{dt} = K_{+E}\,[mRNA \cdot X_F^+] - K_E\,[E \cdot X_F^+] - \frac{D}{\epsilon}\,[E \cdot X_F^+] \quad [8.17]$$

By now familiar equations are used to evaluate the fraction of unbound operator gene, Q, and the binding efficiency, F, of RNA polymerase at the promoter site (Ray et al., 1987).
Operator index:

$$Q = \frac{1 + b_1 \cdot S_{in}}{1 + b_1 \cdot S_{in} + b_2\,[R]_t} = \frac{1 + b_1 \cdot S_{in}}{1 + b_1 \cdot S_{in} + b_2} \quad [8.18]$$

The intracellular lactose concentration, S_{in}, is typically a multiple of the lactose level in the extracellular environment. It may be evaluated with the aid of the following equation.
Intracellular lactose concentration:

$$S_{in} = \frac{A'' \cdot S_F}{B + S_F} \quad [8.19]$$

Catabolic repression index:

$$F = \frac{K_1}{K_2 + [M]} \quad [8.20]$$

The catabolite modulating factor concentration ([M]) has a negative effect on the RNAP binding efficiency. Although its exact chemical structure and mechanism of action are not yet known, it it responsible for the existence of an optimal dilution rate at which the specific enzyme concentration reaches a maximum.   This has been modeled by assuming that a critical growth rate exists such that the concentration of catabolite modulating factor is negligible at growth rates below the critical value, and increases hyperbolically at growth rates above the critical value (Ray, 1985).   Catabolite modulating factor:

$$[M] = \frac{k_m \cdot [\mu_F^+ - \mu_{crit}]}{Y_X \cdot [k_{-m} + (\mu_F^+ - \mu_{crit})]} \qquad [8.21]$$

Lastly, the growth rate of cells in the bulk phase may be evaluated using the following growth model.
Cell growth rate:

$$\mu = \frac{\mu_m \cdot S_F}{S_F + K_s [1 + K_i X]} = \frac{\mu_m \cdot S_F}{S_F + K_s [1 + K_p (S_o - S_F)]} \qquad [8.22]$$

8.10 STEADY STATE IMRC BIOREACTOR DYNAMICS WITH SELECTION PRESSURE

When a sufficient quantity of antibiotic is present in the reactor feed, the growth rate and concentration of plasmid-free cells in the bioreactor is negligible.   For steady state reactor operation, the IMRC reactor balance equations simplify, allowing for an analytical solution of substrate, recombinant cell and enzyme concentrations as functions of dilution rate.   The bulk substrate concentration can be obtained by solving the following quadratic equation.

$$[\nu\delta - \beta] \cdot S_F^2 + [\beta S_o - \kappa - \nu\delta S_o - \nu\lambda] \cdot S_F + \kappa \cdot S_o = 0 \qquad [8.23]$$

where

$$\nu = \frac{\mu_m}{D}$$

$$\beta = 1 - K_s K_p$$

$$\kappa = K_s [1 + K_p S_o]$$

$$\delta = \epsilon [1 - \theta_F]$$

$$\lambda = \frac{\eta_I [1 - \epsilon]}{Y_x} \cdot \frac{[1 - \Theta_I]}{[1 - \Theta_F]} \cdot X_I$$

The substrate concentration is obtained using the quadratic formula; the negative root is the meaningful one.

The free cell growth rate is evaluated using equation [8.22]. The free cell concentration is evaluated using the following equation:

$$X_F^+ = Y_x [1 - \Theta_F] [S_o - S_F] \tag{8.24}$$

The specific activity of β-galactosidase (units/g cells) can be solved directly using the following equation:

$$[E] = \frac{K_{+E} K_{+M} \cdot F \cdot N_p \cdot Q \cdot \mu_F^+}{[K_{-E} + D/\epsilon] \cdot [K_{-M} + D/\epsilon]} \tag{8.25}$$

8.11 ANALYSIS AND DISCUSSION

The IMRC bioreactor model was used to evaluate the experimental results obtained from CSTR and IMRC bioreactor experiments. The bioreactor model was applied to shorter-term steady state bioreactor kinetics using selection pressure (ampicillin), and longer-term transient bioreactor kinetics, in which stability experiments were conducted using a glucose minimal medium. The CSTR bioreactor model is a limiting case of the more general IMRC bioreactor model, and is obtained by setting the liquid fraction (ϵ) equal to one, and simplifying the component balance equations. The steady state bioreactor simulations are another limiting case, and were obtained by setting the time derivatives equal to zero and solving the component balances to obtain cell, substrate and product concentrations.

The parameter values estimated in the following bioreactor simulations were obtained by adjusting them to minimize the sum of squared deviations between the mathematical model and experimental results. This was accomplished through the use of NONLIN, the nonlinear statistical regression package referred to previously.

Mathematical modeling results describing CSTR bioreactor dynamics are shown in Figs. 8.11 to 8.13, along with the experimental

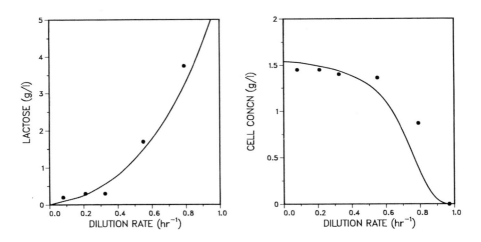

Figure 8.11   Steady state CSTR modeling results: cell and lactose concentration profiles.

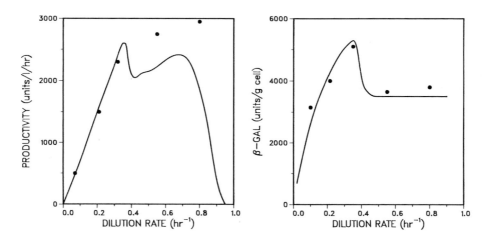

Figure 8.12   Steady state CSTR modeling results: enzyme concentration and productivity
profiles.

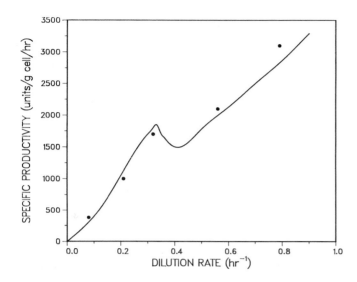

Figure 8.13   Steady state CSTR modeling results: specific productivity profile.

experimental CSTR bioreactor data shown and discussed in a previous section.   Parameter estimates for the biomass yield factor, (Y_x), and the gene expression parameter, b_2 , were obtained using NONLIN to give the best possible fit between the model and experimental data.   The estimate for plasmid copy number N_p was obtained from fitting the unsteady state modeling and experimental results, shown in Figure 8.15.   The parameter value for μ_{crit} was the dilution rate at which the β-galactosidase specific activity displays its maximum value.   The remaining parameter values were taken from previous mathematical modeling studies of wild-type *E. coli*,   carried out in this laboratory (Ray, 1985).   The parameter values used in the simulation are shown in Tables 8.1 and 8.2.

The fit between the mathematical model and experimental data is good and can be attributed in part to the accuracy of the cell growth and lac operon gene expression parameters obtained from previous studies, and also to the the accuracy with which the plasmid copy number could be estimated from the CSTR stability experiment, shown in Figure 8.15. Estimation of one lac operon gene expression parameter value, b_2, was necessary because the plasmid used in this study has a *lac i*^{q} repressor gene which overproduces the lac repressor protein.   The additional level of repression is manifested by increasing the value of this dimensionless parameter, which has an estimated value of 245, as compared to a normal *lac i* gene, which has a value of 100.

Table 8.1   Parameter Values Used in Steady State Bioreactor Model

Parameters	Values	Parameters	Values
μ_m	0.92 hr^{-1}	N_p	7.2 plasmids/cell
μ_{crit}	0.35 hr^{-1}	K_{+E}	1000 units $\cdot mg^{-1} \cdot hr^{-1}$
Y_x	0.32	K_{-E}	0.0 hr^{-1}
K_s	0.04 g/l	b_1	5.0 l/g
K_p	10.0 l/g	b_2	245 dimensionless
K_1	4.82 mg/l	A''	60.0 g/l
K_2	4.89 mg/l	B	0.40 g/l
K_{+M}	50.0 mg/g	k_{+m}	1.48 mg/l
K_{-M}	27.6 hr^{-1}	k_{-m}	0.05 hr^{-1}

Table 8.2   Operational Parameter Values Used in Steady State Bioreactor Simulations

CSTR		IMRC	
S_o	5.0 g/l	S_o	5.0 g/l
ϵ	1.0	ϵ	0.714
		X_I	3.25 g/l
		η_I	0.22

A simulation of steady state IMRC bioreactor kinetics (Fig. 8.14) was performed using the parameter values shown in Tables 8.1 and 8.2. When compared against the IMRC bioreactor experiment (e.g., Figure 8.8), the model predicts higher cell concentrations at low dilution rates than those measured experimentally.   At high dilution rates, the experimental cell concentrations are slightly higher.   The lack of conformity of the experimental data to the model is due to the phosphate limitations discussed earlier.   When compared against the CSTR bioreactor model, the IMRC bioreactor model predicts a maximum enzyme volumetric productivity at a lower dilution rate than that predicted by the CSTR bioreactor model. The IMRC bioreactor model does predict substantial enzyme productivities at dilution rates above the maximum specific growth rate, thus demonstrating the enhanced operational stability of the IMRC bioreactor at high dilution rate. With higher immobilized cell loading or a protein secretion system, the IMRC productivity values could be increased dramatically.

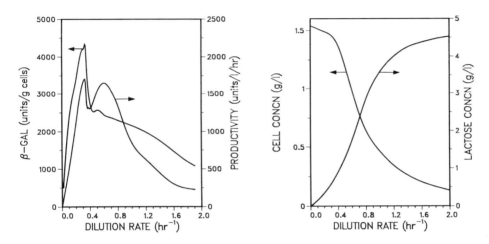

Figure 8.14   Steady state IMRC bioreactor simulation results.

The mathematical modeling of plasmid segregational instability kinetics requires consideration of the cell growth kinetics only. The gene expression kinetics do not enter the calculations, because very little product was produced when the recombinant cells were grown on glucose. The phosphate limitation was obviated by switching to κ-carageenan as the carrier. The focus of this phase of the modeling and experimental study was to analyze the loss of the recombinant cell population from the bioreactor, and the mechanisms involved. There are two main processes which govern the rise in the fraction of wild-type (plasmid-free) cells in the bioreactor. The first is the rate at which recombinant cells divide to give one recombinant and one plasmid-free cell. This process of unequal plasmid partitioning gives an exponential decline in the recombinant cell fraction. The second process involves the ratio of growth rates between the wild-type and recombinant cells, which causes the fraction of recombinant cells to decrease as a result of having a longer doubling time. This growth-rate difference manifests itself more slowly than the plasmid partitioning process, in that it becomes more noticeable as the fraction of wild-type cells increases.

The parameter values used in the simulation, shown in Figure 8.15, are listed in Table 8.3. The parameters estimated in this simulation are the plasmid copy number (N_p), and the ratio of growth rates between the wild-type and recombinant cells (α). The remaining parameter values are operational constants used in the bioreactor experiment.

The estimated value of the mean plasmid copy number, N_p, was 7.2 copies/cell, which is less than the value of 11 obtained by isolation and quantification of plasmid from shake-flask experimentation. The statistical estimate is likely to be the more accurate of the two methods, due to the inaccuracies of gel electrophoresis and cell plating quantification procedures. The estimated value of α was 1.02 which indicates that the plasmid-free cell growth rate is 2% greater than the recombinant cell growth rate, which is a very small difference. Our results for N_p and α are in good agreement with those of Seo and Bailey (1985) and Shuler (1986). Had a parameter value of 1.0 been used for α, the estimated plasmid copy number would have been 6.5 copies/cell. Therefore, in this experiment, the main event governing the rate of decline in the recombinant cell fraction is unequal plasmid partitioning.

Table 8.3   List of Operational and Estimated Parameter Values Used in Bioreactor Stability Simulations

CSTR		IMRC	
$\mu_F^+ = D$	0.67 hr^{-1}	D	1.08 hr^{-1}
ϵ	1.0	ϵ	0.71
X_F^+ (t=0)	1.57 g/l	X_I	3.25 g/l
ϕ_F (t=0)	1.0	N_p	7.2 plasmids/cell
α	1.02	X_F^+ (t=0)	0.35 g/l
N_p	7.2 plasmids/cell	μ_F^-	0.98 hr^{-1}
		α	1.10
		η_I	0.22

The parameters used in the IMRC bioreactor stability simulation (Fig. 8.16) are shown in Table 8.3. The plasmid copy number estimate from the free cell bioreactor stability experiment was used. The value of η_I was estimated by simulation, using the IMRC bioreactor equations, to obtain the observed free cell concentration at the initial conditions. The specific growth rate of wild-type cells (μ_F^-) was obtained by solving the IMRC bioreactor model equations at the dilution rate used in the experiment. The value obtained is the same as the maximum specific growth rate, as a result of the high dilution rate providing ample substrate to the free cells. The ratio of growth rates between wild-type and recombinant cells is very sensitive to the actual growth rate. It has been shown that this ratio increases with decreasing growth rate (Seo and Bailey, 1985). The statistical estimate of α obtained in this simulation was 1.10, which is a significant increase over the value of 1.02 obtained in the free cell bioreactor simulation, shown in Fig. 8.15.

The difference in growth rates between recombinant and wild-type cells becomes very important in an immobilized cell bioreactor system. Despite the fact that this ratio increases the rate of loss of plasmid-containing cells, there is a marked reduction in the overall rate due to the slower growth rate of immobilized cells. It is this net improvement in the time profile for the recombinant cell fraction that suggests that immobilized cell bioreactor systems be tested widely for protein product biosynthesis.

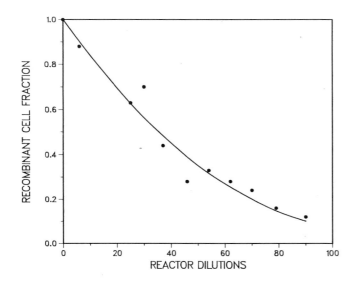

Figure 8.15 CSTR bioreactor stability modeling results.

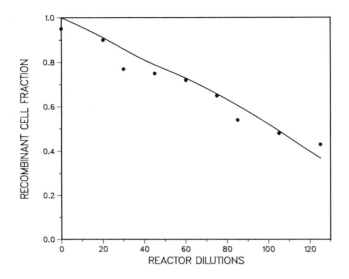

Figure 8.16 IMRC bioreactor stability modeling results.

A total of five parameters were estimated using nonlinear regression. The parameter estimates, standard deviations and 95% confidence intervals are shown in Table 8.4. It is apparent from the values of α obtained in each bioreactor system and the confidence intervals associated with each that a clear difference exists in the ratio of wild-type to recombinant cell growth rates between the two bioreactor systems, and this ratio becomes more important in the process kinetics in the slower-growing immobilized cell system.

Table 8.4   Parameter Estimates and Regression Statistics

Parameter	Estimate	σ	95% C.I.
Y_x	0.32	0.04	0.24 - 0.40
N_p	7.2	0.38	6.35 - 8.07
b_2	244.7	22.7	181 - 308
α_{CSTR}	1.02	0.01	0.99 - 1.05
α_{IMRC}	1.10	0.007	1.09 - 1.12

The single most important conclusion that can be deduced from the experimental results is that IMRC bioreactor systems demonstrate dramatically improved operational stability over continuous free cell bioreactor processes in the absence of selection pressure. Using the plasmid containing cell fraction as the indicator, there is a five to ten fold improvement (i.e., decrease) in the level of plasmid segregational instability in going from the CSTR to the IMRC bioreactor system. This appears to have great potential when applied to a continuous large scale protein production process, where trace quantities of antibiotic in the final product would violate good manufacturing practice (GMP) guidelines, set forth by the Food and Drug Administration.

Mathematical analysis of IMRC and CSTR bioreactor kinetics is an important tool in developing new bioreactor design and operation strategies, as well as in allowing one to evaluate observed experimental results with greater understanding and insight. The most important outcome of the bioreactor simulation efforts was the finding that growth rate differences between wild-type and recombinant cells are magnified in the IMRC bioreactor system, in comparison to the free cell bioreactor system. In addition, application of the plasmid copy number estimate obtained in the free cell bioreactor stability study to steady state gene expression kinetics provided good agreement with the observed levels of enzyme biosynthesis.

The results from CSTR and IMRC bioreactor studies are in agreement with results obtained in a similar CSTR bioreactor study of selection between two wild-type cell populations, in which wall growth (i.e., *self-immobilization*) effects played a significant role in the long-term bioreactor kinetics (Dykhuizen and Hartl, 1983). In our work, the rate of selection decreased significantly in the presence of a fixed (immobilized) cell population. A recent study examining the kinetics of a recombinant *E. coli* strain in CSTR and IMRC bioreactor systems (De Taxis du Poet et al., 1986) also showed similar findings. The study showed a significant decrease in the selection rate between CSTR and IMRC bioreactor systems; however, the link between the observed bioreactor kinetics and longer cell doubling times in the immobilized state awaited our study. The latter has provided a clearer picture of the relative importance of plasmid partitioning kinetics and the ratio of recombinant to free cell growth rates in the overall bioreactor kinetics, using a rigorous bioreactor model and statistical analysis of the parameters contained therein.

Having considered the influence of *intracellular* chemical messenger transport on the regulation of enzyme biosynthesis rates for both wild-type and recombinant cultures, the book now moves on to a consideration of the role of *intercellular* chemical messengers across synaptic junctions.   This subject is taken up in detail in the next chapter.

8.12 NOMENCLATURE (ADDITIONAL TO CHAPTER 7 LIST)

D_{eff}	intraparticle substrate diffusion coefficient, cm^2/sec
$[E]$	β-galactosidase specific activity, units/g cells
$[E \cdot X_F^+]$	β-galactosidase concentration, units/l
G_I	immobilized cell doubling time, hr
m	number of generations
$[mRNA \cdot X_F^+]$	mRNA concentration, mg/l
N_p	plasmid copy number
r	radial distance from particle center, cm
S	substrate concentration, g/l
S_F	liquid phase substrate concentration in the microenvironment of a cell, g/l
S_{in}	intracellular lactose concentration, g/l
S_o	inlet substrate concentration, g/l
X	cell concentration, g/l
Y_x	biomass yield coefficient

SUBSCRIPTS
F	free (liquid) phase
I	immobilized phase

SUPERSCRIPTS
+	recombinant (plasmid-containing)
-	wild-type (plasmid-free)

GREEK SYMBOLS
α	ratio of wild-type to recombinant cell growth rates
ϵ	liquid fraction
η_I	ratio of IMC to free cell growth rate $[\mu_I/\mu_F]$
θ	probability of obtaining plasmid-free segregant
μ	specific growth rate, hr^{-1}
μ_m	maximum specific growth rate, hr^{-1}
σ	standard deviation
ξ_I	substrate diffusion effectiveness factor
ϕ	recombinant cell fraction

8.13 REFERENCES

Bailey, J.E., M. Hjortso and F. Srienc, *Ann. N. Y. Acad. Sci., Biochem. Eng. III*, **413**, 71 (1983).

Bailey, K., Ph.D. Thesis in Chemical and Biochemical Engineering, Rutgers U. (1986).

Berman, M.L. and D.E. Jackson, *J. Bacteriol.*, **159**, 750 (1984).

Casadaban, M.J., *J. Mol. Biol.*, **104**, 541 (1976).

Casadaban, M.J. and S.N. Cohen, *J. Mol. Biol.*, **138**, 179 (1980).

De Crombrugghe, B., S. Busby and H. Buc, *Science*, **224**, 831 (1984).

Dessein, A., F. Tillier and A. Ullmann, *Molec. Gen. Genet.*, **162**, 89 (1978).

De Taxis du Poet, P., P. Dhulster, J.N. Barbotin and D. Thomas, *J. Bacteriol.*, **165**, 871 (1986).

Dhurjati, M., *Biochem. Eng. V Abstracts*, Engineering Foundation Conferences, New York (1986).

Dykhuizen, D.E. and D.L. Hartl, *Microbiol. Rev.*, **47**,150 (1983).

Guidi-Rontani, C., A. Danchin and A. Ullmann, *Proc. Natl. Acad. Sci. USA*, **77**, 5799 (1980).

Karkare, S.B., R.C. Dean and K. Venkatasubramanian, *Biotechnology*, **3**, 247 1985).

Kaushik, K.R., W.R. Vieth and K. Venkatasubramanian, *J. Molec. Catalysis*, **30**, 39 (1985).

Lee, S.B. and J.E. Bailey, *Biotechnol. Bioeng.*, **26**, 1372 (1984).

Lee, S.B. and J.E. Bailey, *Biotechnol. Bioeng.*, **26**, 1383 (1984).

Maniatis, T., E.F. Fritsch and J. Sambrook, *Molecular Cloning*, Cold Spring Harbor (1982).

Messing, J., R. Crea and P.H. Seeburg, *Nucl. Acids Res.*, **9**, 309 (1981).

Ray, N.G., Ph.D. Thesis in Chemical and Biochemical Engineering, Rutgers U. (1985).

Ray, N.G., W.R. Vieth and K. Venkatasubramanian, *Ann. N.Y. Acad. Sci., Biochem. Eng. IV*, **469**, 212 (1986).

Ray, N.G., *Biochem. Eng. V Abstracts*, Engineering Foundation Conferences, New York (1986).

Ray, N.G., W.R. Vieth and K. Venkatasubramanian, *Biotechnol. Bioeng.*, **29**, 1003 (1987).

Ryu, D., *Biochem. Eng. V Abstracts*, Engineering Foundation Conferences, New York (1986).

Scott, J.R., *Microbiol. Rev.*, **48**, 1 (1984).

Seo, J.H. and J.E. Bailey, *Biotechnol. Bioeng.*, **27**, 156 (1985).
Shuler, M., *Biochem.Eng.V Abstracts*, Engineering Foundation
    Conferences, New York (1986).
Vieira, J. and J. Messing, *Gene*, **19**, 259 (1982).
Wanner, B.L., R. Kodaira and F.C. Neidhardt, *J. Bacteriol.*, **136**, 947
    (1978).

I think my song has lasted almost long
    enough.
The subject's interestin', but the rhymes
    are kinda tough.

Folk Ballad, "Goober Peas"

9

INTERCELLULAR CHEMICAL MESSENGER TRANSPORT AND SYNAPTIC RESPONSE

9.0 INTRODUCTION

With an abiding interest in the biopolymer, collagen, our investigations now take us into the realm of biomembrane structures at synaptic junctions. By examining reconstituted biomembrane structures as models of the synaptic cleft-postsynaptic membrane combination, it is possible to estimate the effective diffusivity of the *intercellular* chemical messenger, acetylcholine, in a membrane laminate structure. The next step is to integrate these results with an appropriate description of the accompanying chemical rate processes, to arrive at a model with a predictive capability. As an accompanying motivation, simulations of this kind are expected to be helpful in suggesting technological designs for microscopic, membrane-based chemical sensors of the more advanced type, as depicted in Chapter 4. These points are taken up in the discussion which follows.

9.1 CHEMICAL MESSENGER BINDING AND TRANSPORT IN RECONSTITUTED BIOMEMBRANE STRUCTURES

Recently, it was found that dual sorption processes were applying in sucrose transport in a biopolymeric membrane (i.e., collagen) system. Ludolph et al. (1979) found that such effects extended to enzyme-bound collagen membranes. Soddu and Vieth (1980) identified two types of sites in the collagen structure to which the substrate (sucrose) has different affinities, the substrate fraction bound to the high-affinity sites being immobile. It was hypothesized that this model might be gainfully adapted to describe a number of the observed transport properties of other biological membranous structures. This suggestion leads to a more extensive investigation into the occurrence of transport retardation owing to penetrant immobilization effects in biological membrane systems.

An important representative biological membrane system is the structure comprising the *collagenous* synaptic cleft ending in the receptor-rich post-synaptic membrane.   Binding of acetylcholine to the latter triggers the opening of discrete channels or pores through which sodium and potassium ions counterflow along their electrochemical gradients.   In this biomembrane system, binding phenomena in relation to the nerve mechanism and the roles of several substrates in neurotransmission have been studied in detail (O'Brien, 1980).   For instance,   Heidmann and Changeux (1978) and Couteaux (1958), among many other researchers, have investigated the mechanism of the response of the excitable membrane.   Elliot et al. (1980) and Lindstrom et al.   (1980) have demonstrated successful purification and characterization of the structures of acetylcholine receptor from the electro-organs of torpedo fish and electric eel.

As the first step in a series of research efforts (Hirose and Vieth, 1983), acetylcholine iodide and swollen collagen were selected as models of the penetrant and the fluid-filled collagenous matrix of the synaptic cleft (Lester, 1977), respectively.   The aim was to experimentally characterize the basic phenomenon of the chemical messenger transport process across the synaptic cleft.   Transport and sorption under the conditions of varying acetylcholine halide concentrations were investigated.   In addition, as the beginning of the second phase of work, accumulations of AChI at low bulk concentrations (within the physiological range) in vesicles which comprise ACh receptor-rich membrane were investigated.

9.2 COLLAGEN MICROSTRUCTURE

Collagen is a ubiquitous and effective structural protein in mammalian vertebrate systems.   It is composed of macrofibrillar bundles, themselves made up of microfibrillar elements which, at the molecular level, are referred to as tropocollagen.   Tropocollagen, in turn, comprises association oligomers (e.g., dimers, pentamers) of the individual macromolecular triple helices.   Each helical strand is rich in glycine and hydroxyproline, as well as polar amino acid residues in "bonding zones."

A useful functional model is that proposed by Grant et al. (1967).   They regard tropocollagen as composed of 5 so-called bonding zones (a zones) separated by 4 so-called non-bonding zones (b zones). The model does not require that each zone, even of a given type, have components of the same chemical structure but it does postulate that aggregation is determined by the tendency for the a bonding zones of any one tropocollagen to become associated with like bonding zones on adjacent molecules.   Individual chains may twist and rotate and even cross over one another in order to maximize the bonding.   The model also provides for considerable free volume resulting primarily from the smaller areal density of macromolecules in the aggregated b or non-binding zones.   Thus, the exposed surface area of fibrils is substantial and diffusion of a variety of molecules, such as hormones (e.g., acetylcholine) and enzymes, to appropriate binding sites would be affected.

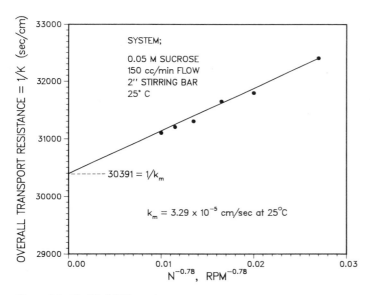

Figure 9.1 Modified Wilson plot.

9.3 TRANSPORT STUDIES

AChI time lag and diffusivity were measured using the same method and apparatus as in the previous study (Ludolph et al., 1979), employing a diffusion cell consisting of two very well stirred compartments to minimize and obviate the bulk phase mass transfer resistance (see Fig. 9.1).   AChI solution recycled upstream while phosphate buffer (0.05 M, pH 7.0) passed over the downstream side.   The total amount (Q_t) of penetrant (AChI) that permeates through the membrane in time (t) can be calculated as follows:

$$Q_t = \int_0^t JAdt = \int_0^t [V \, dc/dt \,]_t + qc \,] dt \qquad [9.1]$$

where J is the flux of AChI, c is the *downstream concentration* of AChI at time t, A is the membrane area, V is the volume of the diffusion cell and q is the flow rate of the phosphate buffer (0.05 M, pH 7.0). The time lag (Θ) was graphically obtained from the plots of Q_t aganst t. The asymptotic time lag, Θ_{asym}, was obtained from plots of Θ against upstream AChI concentration, which displayed a nonlinear transition region asymptotically tending to a constant value in higher AChI concentrations. The diffusivity (D_{asym}) of the penetrant can be determined as follows:

$$D_{asym} = L^2/6\Theta_{asym} \qquad [9.2]$$

where L is swollen membrane thickness.
This asymptotic value (D_{asym}) was then used to calculate mobile species concentrations. At steady state,

$$J = D_{asym} [(C_{MU} - C_{MD})/L] \qquad [9.3]$$

where C_{MU} and C_{MD} are the upstream and downstream concentrations of the mobile species, respectively. In this experiment, the concentration of the mobile species in the downstream side (C_{MD}) is negligible compared to that in the upstream side (C_{MU}). Therefore, eqn. [9.3] can be rewritten as:

$$J = D_{asym} C_{MU}/L \qquad [9.4]$$

C_{MU} is then expressed by:

$$C_{MU} = JL/D_{asym} \qquad [9.5]$$

J can be calculated from the time lag plot. Then, the concentration (C_I) of the immobile species in the membrane is expressed by:

$$C_I = C_T - C_{MU} \qquad [9.6]$$

where C_T is the total concentration obtained from a sorption experiment.

9.4 PURIFICATION AND RECONSTITUTION OF VESICLES FROM TORPEDO FISH

This procedure followed the general approach of several laboratories in which acetylcholine receptor (AChR)-rich membrane vesicles were purified. Because they may be a bit unfamiliar to some readers, the method of preparation and the assay are outlined briefly as follows:

After *Torpedo California* electroplax organs were dissected away, the organ (ca. 100 g) was minced into small pieces and an equal volume of cold buffer (10 mM sodium phosphate pH 7.8, 400 mM NaCl, 5 mM EDTA, 0.02% NaN_3, 5 mM iodoacetamide) was added. After 2 min initial grind at high speed in a commercial Waring blender, the homogenate was reground in small portions for four periods of 30 s at 30,000 rev/min in a Virtis 60. Connective tissue and other large particles were pelleted by centrifugation of the homogenate at 5,000 rev/min for 10 min in a Sorvall GS-2 at 2°C. The supernatant was passed through two layers of cheesecloth and centrifuged at 16,000 rev/min for 1 hr in a Sorvall GS-2 rotor at 2°C. The pellet obtained was comprised of AChR-rich membrane vesicles to be further reconstituted by soybean lipids. Vesicles were resuspended in reconstitution buffer (2% Na cholate (Sigma); 25 mg/ml of soybean L-α-phosphatidylcholine (Sigma); 100 mM NaCl; 10 mM phosphate buffer, pH 7.5; 10 mM NaN_3) at concentrations of 0.02 to 0.1 g of vesicles/ml. After blending on a Vortex mixer and stirring to provide a homogeneous suspension, this mixture was dialyzed overnight against at least 100 volumes of 100 mM NaCl, 10 mM Na phosphate, pH 7.5, and 10 mM NaN_3, followed by an additional overnight dialysis against flux buffer (145 mM sucrose, 10 mM Na phosphate, pH 7.5, 5 mM NaN_3).

9.5 CARBAMYLCHOLINE-INDUCED $^{22}NA^+$ AND ACHI UPTAKES OF RECONSTITUTED VESICLES

$^{22}Na^+$ uptake was conducted at room temperature behind suitable lead shielding. In microfuge tubes, $^{22}Na^+$ (5μl of 0.2 mCi/ml) and water (5μl) or carbamylcholine (5 μl of 1 x 10^{-4} M) were mixed. The assay was initiated by pipetting in 40 μl of vesicles in flux buffer with an Eppendorf pipetter (Lindstrom et al., 1980). After mixing, the mixture was transferred to a column (0.5 x 8 cm) packed with Dowex 50W-X8-100 (Sigma) treated with Trizma base (Sigma); then, sucrose solution (3 ml at 175 mM) was carefully added to elute the vesicles. The tube containing the eluate of each column was placed in the γ-counter and counted. The carbamylcholine-induced $^{22}Na^+$ uptake was defined as the difference between the counts per minute in the water samples and those containing carbamylcholine.

The AChI uptake of reconstituted vesicles was carried out by a method very similar to the one given above. That is, in microfuge tubes, AChI in the range from 10^{-6} to 10^{-1} M (450 μl), radioactive AChI (30 μl 0.01 mCi/ml) and reconstituted vesicles (20 μl) were mixed. The mixture was transferred to the column (0.5 x 8 cm) and sucrose solution (3 ml at 175 mM) was carefully added to elute the vesicles. The eluate (0.5 ml) was mixed into liquid scintillation cocktail (10 ml) and AChI concentrations in vesicles were calculated.

9.6 ISOTHERM RESOLUTION

In order to measure the effect of AChI concentration on the swelling of collagen, membrane samples were swollen at ph 7.0 (phosphate buffer, 0.05 M). In Fig. 9.2, the experimental data are plotted in the range extending to 2.0 mol/l of AChI, for a collagen membrane of initial swelling ratio 16.6. This ratio remains constant, until, at AChI concentrations of ca. 2.0 mol/l, there is an abrupt change in swelling. It was noticed that immersion of the collagen membranes at high AChI concentrations resulted in shrinkage due to interaction between AChI and collagen. This type of result was previously observed with sucrose and collagen by Soddu and Vieth (1980).

AChI sorption in collagen membranes was studied over a wide range of concentrations and degrees of swelling. For these experiments, collagen membranes having various degrees of dehydrothermal crosslinking were prepared as summarized in Table 9.1. The results of sorption experiments are shown in Fig. 9.3. Large differences in the levels of sorption were found for different degrees of swelling. For high degrees of swelling, the level of sorption proportionally increased with increasing AChI concentrations. For swelling ratios 9.9, 13.4 and 16.6, AChI was slightly positively accumulated in the collagen membrane compared with bulk AChI concentrations. For low degrees of swelling, 6.3 and 3.1, which were obtained by a long-term aging of collagen and the reaction of glutaraldehyde, respectively, the slope reached a plateau at AChI concentrations higher than 1.5 mol/l.

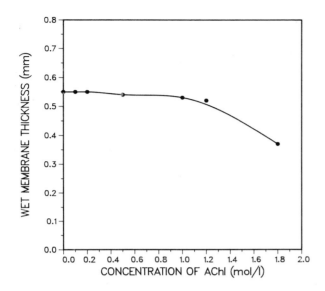

Figure 9.2   Swelling of collagen membrane at pH 7.0; degree of swelling, 16.6.

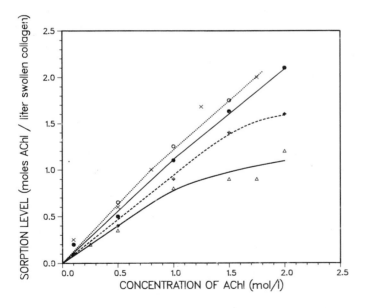

Figure 9.3   Sorption level at different degrees of swelling: (X) 16.6, (●) 13.4, (O) 9.9,   (▲) 6.3 and (Δ) 3.1.

For high degrees of swelling, the fluid-filled structure is highly open and all microdomains are accessible to the substrate; AChI occupies affinity sites of the collagen macromolecules. In contrast, relative inaccessibility of microdomains (Grant et al., 1967) is the limitation of highly crosslinked structures at low degrees of swelling. Nonetheless, a substantial fraction of the microdomains is still accessible to the substrate even in a highly crosslinked structure. Therefore, saturation effects of sorption level were observed at higher AChI concentrations.

Table 9.1   Dehydrothermal Crosslinking of Collagen Membranes

Swelling ratio	Swollen membrane thickness (μm)	Aging conditions
16.6	830	4 days at 25°C
13.4	1075	4 days at 25°C
9.9	893	31 days at 25°C
6.3	506	10 days at 65°C and 90% rel.humidity
3.1	301	(5 min in 3% (v/v) glutaraldehyde sol.)

Time lag experiments were performed using a collagen membrane of degree of swelling 16.6. The values of Θ are plotted in Fig. 9.4. The time lag asymptotically decreased with increasing upstream AChI concentrations, following the approach to Θ_{asym}=2.0 min. This type of behavior is predicted by the analysis of Bhatia and Vieth (1980).

Next, AChI concentrations of the mobile species (C_{MU}) were calculated from the flux and the asymptotic diffusivity, for degrees of swelling 6.3, 13.4 and 16.6. The AChI concentrations (C_I) of the immobile species were also calculated as the difference between the value of the total sorption level of AChI and the concentrations of the mobile species, using eqn. [9.6]. The data are illustrated in Figs. 9.5, 9.6 and 9.7. These results are also in agreement with the experimental findings of other researchers (Weadock, 1986; Weadock et al., 1986).

The mobile species concentration increased with increasing upstream AChI concentration, while it was roughly independent of the degree of swelling. Sigmoidal curves for the immobile species were obtained for all degrees of swelling, characteristic of an allosteric transport effect (Soddu and Vieth, 1980). It should be noted that the

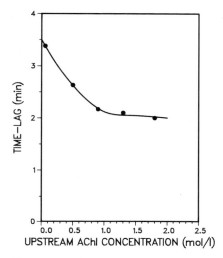

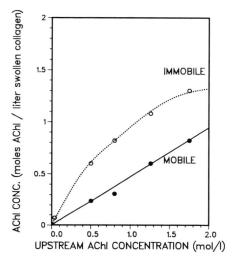

Figure 9.4   Relation between time lags and upstream AChI concentrations with collagen membrane; degree of swelling 16.6.

Figure 9.5   Resolution of AChI concentrations of mobile and immobile species in collagen membrane at 25°C; degree of swelling, 16.6.

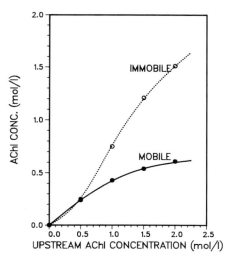

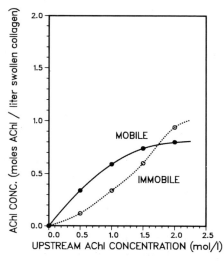

Figure 9.6   Resolution of AChI concentrations of mobile and immobile species in collagen membrane at 25°C; degree of swelling, 13.4.

Figure 9.7   Resolution of AChI concentrations of mobile and immobile species in collagen membrane at 25°C; degree of swelling, 6.3.

initial slopes of the sorption curves decreased with decreasing degrees of swelling.   Indeed, the immobile species concentrations decreased remarkably with decreasing degrees of swelling, in contrast to the behavior of the mobile species population.

9.7 VESICLES

It is well-known that cation channel activity of AChR-rich membrane vesicles is dependent on many factors; for instance, the type of purification steps, conditions of vesicle reconstitution, concentrations of inducing carbamylcholine, as well as general environmental conditions, such as pH, ionic strength and temperature (Lindstrom et al., 1980).   In particular, the balances of vesicle and soybean lipid concentrations in reconstitution buffer solution have large effects on the survival of cation channel activities.

In order to prepare an appropriate recipe for reconstituted AChR-rich membrane vesicles, carbamylcholine (10^{-4} M)-induced $^{22}Na^+$ uptakes were studied, under the conditions of different vesicle concentrations, following dilution with flux buffer.   Figure 9.8   shows that carbamylcholine-induced $^{22}Na^+$ uptake increased with increasing vesicle concentrations.   From previous experiments in our laboratory and those of Lindstrom (1980), no cation channel activity survived without addition of lipids into reconstitution buffer. From Anholt's results (1980) on the relation between Na uptake and AChR concentrations, a value of AChR concentration, approximately 0.1 μM, was estimated under the conditions of 0.10 g of vesicles per g reconstitution buffer.

In the range from 10^{-6} to 10^{-1} mol/l of immersing AChI concentrations, sorption levels in collagen specimens with degree of swelling 16.6, were measured.   AChI uptakes of reconstituted vesicles (0.02 g/g) shown in Fig. 9.8 were also measured.   The distribution coefficient, or K value, is defined as the ratio of sorption level (AChI uptake) and bulk AChI concentration; it is shown in Fig. 9.9.   It is of interest that the AChI concentration in collagen membrane was accumulated at least ten-fold in very dilute solution, while nearly one thousand-fold in vesicles.   For the latter case, it is possible to estimate the value of an apparent Michaelis constant for the receptor.   We obtain 2.0×10^{-4} mol/l, in good agreement with the estimated value of ACh concentration in the synaptic cleft during the process of neurotransmission (Heidmann and Changeux, 1978).

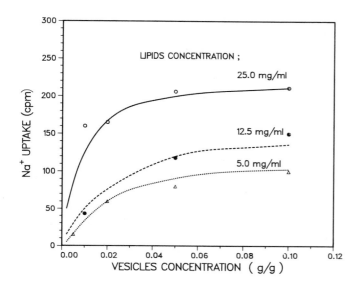

Figure 9.8   Effect of reconstituted vesicle concentration on uptake of ^{22}Na$^+$ in response to 10^{-4} M carbamylcholine.

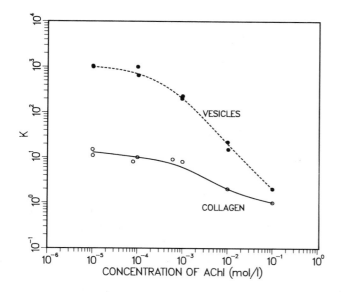

Figure 9.9   Distribution coefficients vs. AChI concentrations for collagen membrane (DS=16.6) and reconstituted AChR-rich vesicles.

9.8 EFFECTIVE DIFFUSIVITY AND LIGAND BINDING EFFECTS

As already described, resolutions of sorption level for various collagen membranes of different degrees of swelling were accomplished with the aid of diffusion experiments. Recalling Fig. 2.11, it is clear that the profile of effective diffusivity displays both higher and lower asymptotic branches. The lower branch is the one which would apply in the dilute physiological concentration range. Therefore, an effective ACh diffusion coefficient, D_{eff}, can be estimated at a value of 1.0×10^{-6} cm^2/sec (see Table 9.2) which is in good agreement with the value found when sucrose was transported through collagen membranes.

Table 9.2   Diffusivities of Acetylcholine in a Swollen Collagen Membrane   (Swelling Ratio = 13.4)

AChI mol/L	θ (Time Lag) Min.	D cm^2/sec
1.780	2.0	3.8×10^{-6} *
1.281	2.2	3.5×10^{-6}
0.823	2.2	3.5×10^{-6}
0.482	2.6	3.2×10^{-6}
0.0965	3.2	2.4×10^{-6}
Infinite Dilution	-----	(1.0×10^{-6}) **

* D = Upper asymptotic value
** D_{eff} = $D_{upper\ asymptote}$ x mobile species fraction at infinite dilution
    = 3.8×10^{-6} x C_M/C_T = 3.8×10^{-6} x 0.25 = $0.95 \times 10^{-6} cm^2/sec$
    where C_M/C_T is estimated from data in Fig. 9.6.

9.9 MEMBRANE LAMINATE MODEL OF THE NEUROMUSCULAR JUNCTION

One of the essential biological membranous systems in animals modulates the transmission of nerve impulses at synapses or neuroeffector junctions. The fusion of vesicles impregnated with acetylcholine with the presynaptic membrane brings about ACh release into the fluid-filled cleft between the terminal and muscle cell. ACh chemical messengers penetrate the synaptic cleft and adhere to receptors embedded in the muscle-cell membrane. Conformational shifts occur, opening channels, allowing both sodium and potassium ion transport through the membrane cation channels.

In order to demonstrate the transport characteristics of model membranes for the synaptic cleft and ACh receptor (AChR) toward neurotransmitters such as ACh, various immobilized preparations of vesicles in liquid membranes laminated to collagen membrane supporting structures were next constructed (Hirose and Vieth, 1984). Following suitable analysis, transport of ACh in synaptic model membranes could be isolated, elucidating the behavior of AChR in retardation of the overall transport process.

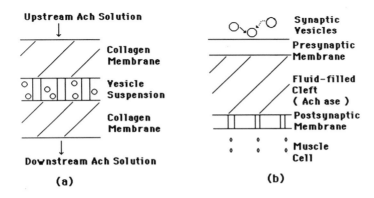

Figure 9.10   Schematic diagram of the liquid membrane system (a) and postsynaptic membrane (b).

Properties of vesicles were routinely assayed, in order to prepare a model of the postsynaptic membrane that comprised AChR impregnated as a liquid membrane into a microporous cellulose host membrane. Freshly prepared vesicles were immediately immobilized as liquid membranes (Fig. 9.10). These membranes were laminated to collagen membranes on one or both sides and joined together to model the cleft and postsynaptic membrane combination. In this manner, the liquid membrane system was made available for investigation of transport of ACh. The time lags (θ_{121}) of the laminate membrane systems were measured under conditions of various ACh concentrations and various volume fractions of vesicles.

In order to estimate the diffusion rate of neurotransmitter in the postsynaptic membrane, transport of ACh through the liquid membrane model was isolated.   The effects of ACh concentrations in upstream solutions on time lags in the transport process, under the conditions of constant volume fraction of vesicle suspensions (0.05), collagen membrane thickness (ca. 150 μm) and temperature (3.0°C) are shown in Fig. 9.11.   The time lags asymptotically decreased with increasing upstream ACh concentrations, approaching 24 minutes when using vesicles and 20 minutes for the vesicle-free control.   Experiments examining the effect of vesicle volume fraction were carried out under standard conditions (10^{-3} mol/l ACh concentration, 3.0°C temperature, ca. 150 μm collagen membrane thickness, 100μm microporous cellulose membrane (5 μm pore size), and a range of vesicle volume fractions from 0.02 to 0.20).   Time lags gradually increased with increasing volume fractions of vesicles.   They were resolved analytically to isolate the effect of vesicle binding on retardation of the transport process.

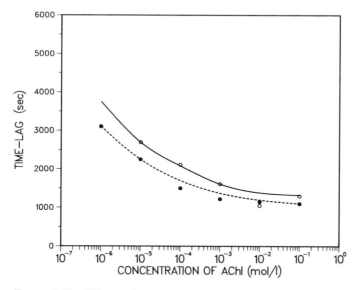

Figure 9.11   Effects of upstream ACh concentrations on time lags in the liquid membrane system: (O) vesicle suspensions in the liquid layer, with volume fraction, 0.05;   (●) flux buffer in the liquid layer.

In general, for a three-layer composite (1,2,1),

$$\theta_{121} = \frac{\dfrac{l_1^2}{D_1} - [4l_1/3D_1K_1 + l_2/D_2K_2] + \dfrac{l_2^2}{D_2}[l_1/D_1K_1 + l_2/6D_2K_2] + \dfrac{l_2 l_1^2 K_2}{[D_1K_1]^2}}{\dfrac{2l_1}{D_1K_1} + \dfrac{l_2}{D_2K_2}} \qquad [9.7]$$

where θ_{121} denotes the time lag of the three layer composite, and l_1, D_1 and K_1 are the thickness, diffusivity and distribution coefficient of the first layer, respectively (Barrie et al., 1963). Subscripts denote the kind of membrane, that is, whether a collagen layer (1) or a vesicle suspension layer (2). For the collagen layer it has been shown that $D_1 = 1.0 \times 10^{-6}$ cm²/s; K_1 is likewise obtained from previous work (Hirose et al., 1983).

Table 9.3   Trend of Diffusivities of ACh in the Vesicle
             Suspension Layer

ACh mol/l	Vol. Fraction of Vesicles, ϵ	K_v	K_1	K_2	$D_2 \times 10^7$ cm²/sec
10^{-3}	0	---	8	1.0	50
10^{-3}	0.02	250	8	6.0	5.5
10^{-3}	0.025	250	8	7.25	4.1
10^{-3}	0.05 *	250	8	13.5	3.2
10^{-3}	0.05 *	250	8	13.5	3.2
10^{-3}	0.10	250	8	26	5.8
10^{-3}	0.20	250	8	51	1.5
10^{-4}	0.05	750	10	38.5	1.6
10^{-5}	0.05	1000	12	51	1.5

*   replicates

For the distribution coefficient of the vesicle suspension layer, K_2,

$$K_2 = ([1 - \epsilon] + \epsilon K_v)/([1 - \epsilon] + \epsilon) = 1 + \epsilon [K_v - 1] \qquad [9.8]$$

where K_v is the distribution coefficient of pelleted vesicles (Hirose et al., 1983) and ϵ is the volume fraction of pelleted vesicles in the vesicle suspension layer.   Experimental data for l_1, l_2 and ϵ were obtained for the laminated structures; values of D_2 could then be calculated from eqn. [9.7].   The diffusivities in vesicle suspension layers were found to decrease with decreasing ACh concentrations and increasing volume fractions of vesicles, showing the by now familiar "duality signature" (Table 9.3).

The neurotransmission system for the cleft plus postsynaptic membrane combination is, of course, considered to be a *two-layer* composite.   Thus,

$$l_{12}/D_{12}\,K_{12} = l_1/D_1\,K_1 + l_2/D_2\,K_2 \qquad [9.9]$$

where

$$K_{12} = [l_1\,K_1 + l_2\,K_2]/l_{12} \qquad [9.10]$$

And, D_1/D_{12} can be written as

$$\frac{D_1}{D_{12}} = \frac{l_1}{l_{12}^2}\,[\,l_1 + l_2\,(K_2/K_1)\,] + \frac{l_2}{l_{12}^2}\,\frac{D_1}{D_2}\,[\,l_1\,(K_1/K_2) + l_2\,] \qquad [9.11]$$

Taking $K_2 = K_v$ at $\epsilon = 1.0$, and D_1 and K_1, 10^{-6} cm^2/s and 10, respectively (Hirose et al., 1983), D_1/D_{12} can be calculated, for the case where mean values of l_1 and l_2 are taken as 900Å and 100Å, respectively (Lester, 1977).   Under the prevailing conditions, i.e., $D_2 < D_1$ and $l_2 < l_1$, eqn. [9.11] simplifies to:

$$D_{12} = \frac{D_1}{1 + \dfrac{l_2\,K_2}{l_1\,K_1}} = \frac{1 \times 10^{-6}}{1 + \dfrac{1}{9}\cdot\dfrac{10^3}{10}} = 8.3 \times 10^{-8}\ cm^2/sec \qquad [9.12]$$

Therefore, an estimate of the order magnitude of the characteristic time scaling parameter for the synaptic cleft-post synaptic membrane combination would yield:

$$\frac{D_{eff}}{l^2} = \frac{8.3 \times 10^{-8}\ cm^2/sec \times sec/10^3\,msec}{10^{-10}\ cm^2} \qquad [9.13]$$

$$= 0.83 \text{ msec}^{-1}$$

As will be shown a bit further along in this chapter, this parameter was independently estimated at a value of 0.84 from resolution of membrane conductivity data.

This level of agreement is really better than one has a right to expect. Nonetheless, these results clearly show the retardation of the diffusion process $(D_{12} < D_1)$ that occurs because a fraction of the penetrant population is immobilized $(K_2 \gg K_1)$ and unavailable for diffusion at any instant. This effect will be only partially offset by the gradient-sharpening effect of the enzyme reaction which will be considered in the next section.

The rise time of the membrane potential in the neurotransmission process is known to be about 100 μs from experimental data. Previous calculations had indicated that the process would require only 20 μs, considering diffusion in the cleft alone (Eccles and Jaeger, 1958). Thus, it was concluded that the diffusion process in the cleft was not controlling. However, for the composite structure of the cleft plus postsynaptic membrane, the effective diffusivity, D_{12}, is substantially reduced below D_1 because of retardation caused by receptor binding. Therefore, our results suggest that the neurotransmission process for ACh can, in fact, be substantially influenced by a modified diffusion process in the laminate structure.

9.10 CHOLINERGIC RECEPTORS INVOLVED IN SYNAPTIC TRANSMISSION

An action potential that invades the presynaptic terminal of a neuron activates voltage-gated calcium channels. The resulting inflow of calcium allows synaptic vesicles, each containing several thousand molecules of transmitter, to bind to specialized release sites called active zones. Intracellular cAMP then helps break open the vesicles. Once released, the transmitter diffuses across the synaptic cleft where it binds to a receptor and opens (or closes) chemically-gated channels, initiating current flow in the postsynaptic cell. Depending on the channels gated by the transmitter, the current will produce excitation or inhibition.

Excitatory transmitters operate by simultaneously increasing permeability to sodium and potassium, while the permeability to potassium and chloride is increased at the inhibitory ones. Most neurons have numerous synaptic inputs, some excitatory and some inhibitory. Characteristically these release small amounts of transmitter to create excitatory postsynaptic potentials (EPSPs) and inhibitory postsynaptic potentials (IPSPs). The impulse generated in the postsynaptic neuron depends on the balance of these influences.

The biosynthesis of ACh is catalyzed by cholinesterase and readily reversed by acetylcholinesterase. The most widely recognized action sites for ACh are the autonomic preganglionics, and the parasympathetic postganglionics. ACh also appears to be the transmitter of motor nerves in all advanced invertebrate animals except the arthropods, where glutamate serves as the excitatory transmitter.

The mammalian neuromuscular junction (see Fig. 9.12) contains the nicotinic ACh receptors whose subunit organization consists of four glycosylated polypeptide chains with apparent molecular weights of 40,000 (α), 50,000 (β), 60,000 (γ), and 65,000 (δ) in an $\alpha_2 \beta \gamma \delta$ stoichiometry (Reynolds & Karlin, 1978). Since the AChR formed by the four subunits contains both the binding sites for cholinergic ligands and the cation gating unit, it is a transmembrane protein. The AChR possesses properties characteristic of typical allosteric proteins, directly relevant to its physiological role as a chemoelectrical transducer. The two regulatory sites that bind ACh are carried by the two, non-neighboring α-subunits and are exposed to the synaptic cleft.

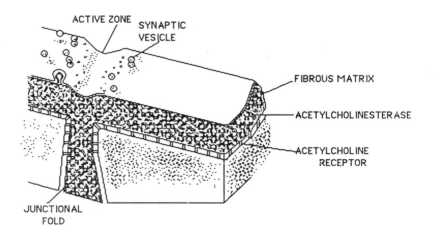

Figure 9.12 Synaptic junction.

The α_2 β $\gamma\delta$ oligomer contains the ion channel, but the precise location is still uncertain. The central axial tunnel might be part of the channel. Most likely, the unique and multichain high-affinity site for non-competitive blockers lies at its level. In addition, multiple low-affinity sites for the ligands are present on the receptor molecule and are most likely located at the interface with the lipid bilayer. These sites and the ACh sites interact with each other in an allosteric, heterotropic manner (Changeux et al., 1983).

There is no strong evidence as to whether the mammalian nicotinic cholinergic postsynaptic membrane possesses separate ionic channels for sodium, potassium and calcium ions except the passage provided by the AChR molecule. The enzyme sodium-potassium-ATPase that serves as the sodium/potassium pump by catalyzing the breakdown of ATP and the calcium/sodium pump that transports one calcium ion outward for every 3 sodium ions transported inward are known to exist in the nicotinic cholinergic synapse, as well as in the muscarinic cholinergic synapse.

9.11 MEMBRANE TRANSPORT /REACTION PROCESSES AT THE NERVE-MUSCLE JUNCTION: NONEQUILIBRIUM MODEL

A uniquely detailed picture of the subtle and intricate pattern of phenomenological interactions governing the response to acetylcholine at the nerve-muscle junction has gradually emerged (e.g., Wathey et al., 1979; Adams et al., 1980; Rosenberry, 1979; Heidmann and Changeux, 1978). At this relatively early stage, however, quantitative description of the phenomenon amounts to a partially solved problem which demands further effort. Progress could conceivably lead to a better understanding of neurotransmitter-receptor interactions, generally.

Recent efforts aimed at quantitative descriptions of the neurotransmission process have attempted to isolate and model a single event that starts in the cleft and concludes in the post-synaptic membrane following release of ACh (Wathey et al., 1979; Adams et al., 1980; Rosenberry, 1979). The common postulation is diffusive radial spreading away from a single vesicle, followed by binding of transmitter in a disc-shaped cleft. However, effects of synchronous and asynchronous release by other vesicles in the local colony which may number as many as several hundred, as described by Kuffler and Yoshikami (1975), had not been analyzed and, on the basis of overly optimistic estimates of the effective diffusivity, the transmitter was considered to appear instantaneously in the cleft. The latter assumption is a fundamental limitation which affects all subsequent estimates of diffusion times, etc. Under these assumptions, several authors have acknowledged that these mathematical characterizations of synaptic transmission ought to be regarded as tentative (Adams et al., 1980; Rosenberry, 1979). To explore another alternative, we have constructed *process* models of continuous and intermittent types, in contrast to *event* type models.

The synaptic cleft is recognized as a fluid-filled fibrous matrix containing immobilized AChase, terminating in a layer of ACh receptors. The matrix itself is composed of collagen and mucopoly-saccharide fibers in association with enzyme. As discussed in the previous section, a number of workers in our laboratory have investigated reconstituted collagen as a carrier for enzymes and have measured and analyzed penetrant transport (e.g., ACh) in such enzyme membrane structures (Soddu and Vieth, 1980; Hirose et al., 1983). We drew upon this experience in approaching the present work. Substantial immobilization of the penetrant population (ca. 75%) has been observed, as described earlier in this chapter (Table 9.2).

9.12 DIFFUSION /ENZYME KINETIC MODEL

Likewise, the analysis of anistropic enzyme membrane structures (biosensors) described in Chapter 4 was helpful in approaching the modeling effort undertaken here. The synaptic cleft is represented as an infinite thin slab or membrane-like diffusion medium, with AChase distributed uniformly within and/or at the post-synaptic surface, which reduces the penetrant time lag. Receptors are located anisotropically at the post-synaptic membrane ($X \equiv 1.0$).

Then, the transient diffusion equation for ACh takes the form:

$$\frac{\partial Y_M}{\partial \theta} = \frac{\partial^2 Y_M}{\partial X^2} - \mu Y_M \qquad [9.14]$$

$$\theta = Dt/l^2 \qquad X = x/l \qquad Y_M = S/S_0$$

The membrane thickness is l and D is the diffusion coefficient. S_0 is the upstream concentration of ACh, Y_M is the dimensionless membrane substrate concentration and μ is a reaction modulus, defined as kl^2/D, where k is the first-order enzyme reaction velocity constant.

The boundary conditions for a unit pulse which commences at $\theta = 0$ and is cut off at $\theta_c = 1.0$ are:

$$Y_M[X, 0] = 0$$

$$Y_M[0, \theta] = 1 \qquad\qquad 0 \leqslant \theta \leqslant \theta_c \qquad\qquad [9.15]$$

$$\qquad = 0 \qquad\qquad \theta \geqslant \theta_c \qquad\qquad [9.16]$$

$$\frac{\partial Y_M}{\partial X} = -\mu Y_M \qquad\qquad X = 1, \theta > 0 \qquad\qquad [9.17]$$

For the enzymatically anisotropic structure of the cleft, eqn. [9.14] becomes a simple diffusion equation,

$$\frac{\partial Y_M}{\partial \theta} = \frac{\partial^2 Y_M}{\partial X^2} \qquad\qquad\qquad [9.18]$$

with hydrolysis described by eqn. [9.17].

Equation [9.14] can be solved mathematically along with eqns. [9.15] - [9.17] by a combination of separation of variables and convolution techniques, to obtain analytical solutions as follows: At $X = 1.0$,

$$Y_M = [\cosh \sqrt{\mu} + \sqrt{\mu} \sinh \sqrt{\mu}]^{-1} + 2 \sum_{n=1}^{\infty} \frac{\sin \lambda_n [\lambda_n^2/\delta_n^2]}{\cos \lambda_n \sin \lambda_n - \lambda_n} \exp[-\delta_n^2 \theta]$$

$$\theta \leqslant \theta_c \qquad\qquad [9.19]$$

$$Y_M = 2 \sum_{n=1}^{\infty} \frac{\sin \lambda_n [\lambda_n^2/\delta_n^2]}{\lambda_n - \cos \lambda_n \sin \lambda_n} [\exp\{-\delta_n^2[\theta - \theta_c]\} - \exp[-\delta_n^2 \theta]\}$$

$$\theta \geqslant \theta_c \qquad\qquad [9.20]$$

where

$$\lambda_n \cos \lambda_n = -\mu \sin \lambda_n \text{ (eigenvalues)} \qquad\qquad [9.21]$$

and $\delta_n{}^2 = \lambda_n{}^2 + \mu$

The enzyme reaction is modeled as a first-order process. This is in agreement with the information that the ACh concentration in the cleft is at about the level of the Michaelis constant, K_M, for AChase (Heidmann and Changeux, 1978; Barman, 1969).

9.13 CONSIDERATION OF OTHER GEOMETRIES

It is not difficult to generalize the solutions appearing in equations [9.19] - [9.21]. Carslaw and Jaeger (1959) provide product solutions for the temperature field v(r, ϕ, z) in the analogous problem of heat transfer in a finite cylinder $(0 \leqslant r \leqslant a, -l < z < l)$.

$$\frac{\partial v}{\partial t} = K \nabla^2 v \tag{9.22}$$

$$v = f[r, \phi, z] \quad \text{when} \quad t = 0 \tag{9.23}$$

$$v = 0 \quad \text{when} \quad r = a \text{ and } z = \pm l \tag{9.24}$$

The solutions have the form,

$$v = \sum_q \sum_{m=1}^{\infty} \sum_{n=0}^{\infty} e^{-k[q^2+m^2\pi^2/4l^2]t} J_n[qr] \sin\frac{m\pi[z+l]}{2l}[A_{q,m,n}\cos n\phi + B_{q,m,n}\sin\phi] \tag{9.25}$$

The point is that, if radial effects on *time delays* do come into the picture, they can be handled simply by inclusion of another characteristic term in the decay constant. Therefore, when in eqn. [9.19] we observe the term, $\exp.(-\delta_n{}^2\Theta)$, we are looking at:

$$\exp. [-D [\delta^2{}_n / L^2] t]$$

If we wish to allow for possible radial effects we may add, in suitable form, a term of the type q^2 where q is a root of $J_n(q,a) = 0$ (Carslaw and Jaeger, 1959), to obtain:

$$\exp. [-D [\delta^2{}_n / L^2 + q^2] t]$$

Therefore, if the term $\delta_n^2 \cdot D / L^2$ appears to be of insufficient magnitude as a predictor for a decay constant, and we wish to explore radial geometry, it is valid to adjust its magnitude by $q^2 \cdot D$.

On the other hand, if we imagine the cleft to be a thin finite rectangle extending indefinitely in one dimension, the solution (Carslaw and Jaeger, 1959) is:

$$v = \psi \, [x, L] \, \psi \, [y, b] \qquad\qquad [9.26]$$

and the decay constant will be modified to a form such as:

$$\exp. \, [\, -D \, [\, \delta_n^2 / L^2 + p^2 / b^2 \,] \, t \,]$$

9.14 ENDPLATE POTENTIAL: EXCITATION AND CONDUCTION OF IMPULSES IN NERVES

The endplate system is considered to transmit signals by nerve pulses called endplate potentials (epp). Several characteristic features of such nerve pulses are: each rises to a level of ca. 20 mV in amplitude above the rest state, and is of the order of msec in duration, and pulses appear to be similar in shape. *The crucial observation is that the transport of sodium takes place in the form of a rapid spike which initiates the rise of the potential, while the counterflow of potassium takes place much more slowly and follows the potential.*

A continuous process of release of transmitter is an approximation of the quantal release pattern and it is the number of released ACh quanta of unit size which dictate the size of the epp (Katz, 1966). The released transmitter diffuses across the synaptic cleft, reacts with receptor molecules in the post-synaptic membrane and alters the properties of the membrane. According to a simple picture of the ion channels, complex protein molecules (receptors) embedded in the membrane allow permeation of small cations such as sodium, potassium, calcium, etc. between the inside and outside of the membrane after transmitter-induced spatial relocation of a charged region (Ehrenstein, 1976). Under the effect, sodium ions flow across the post-synaptic membrane from an outside pool at high concentration and cause a further depolarization, thereby bringing about peaking of the endplate potential. At this point, the potassium flux rises so that potassium ions leave the synapse at a higher rate than that of the entering sodium ions. This brings the membrane back to its null potential. Under short time stimulus, the net membrane current density is approximately zero since capacity current balances out the potassium and sodium currents (assuming there is no leakage current involved). Therefore the equation (Hodgkin and Huxley, 1952) for the membrane potential is:

$$- C_M \frac{dV}{dt} = g_K [V - V_K] + g_{Na} [V - V_{Na}]$$ [9.28]

This equation has been solved under the boundary condition:

$$t = 0, \quad V = \text{rest potential}$$

where V = membrane potential, mv,
V_{Na}, V_K = Nernst potentials, mv,
C_M = membrane capacitance, $\mu F/cm^2$,
g_{Na}, g_K = conductivities for Na, K, mmho/cm^2.
Equation [9.28] can be put into a more convenient form, as follows. Let:

$$\hat{V} = V - \frac{g_K V_K + g_{Na} V_{Na}}{g_K + g_{Na}} \qquad \lambda^2 = \frac{K C_M}{g_K + g_{Na}} \qquad \tau_m = \frac{C_M}{g_K + g_{Na}}$$ [9.29]

Then,

$$- \lambda^2 \frac{d^2 \hat{V}}{dx^2} + \tau_m \frac{d\hat{V}}{dt} + \hat{V} = 0$$ [9.30]

At the membrane, i.e., x = 0

$$\tau_m \frac{d\hat{V}}{dt} + \hat{V} = 0$$ [9.31]

Over a small interval of time,

$$\hat{V} = \hat{V}_0 \exp [- t / \tau_m]$$ [9.32]

where

$$\hat{V}_0 = - \frac{g_K V_K + g_{Na} V_{Na}}{g_K + g_{Na}}$$

and τ_m is a function of time because of its dependence on $g_K(t)$, $g_{Na}(t)$.

Equation [9.31] can be solved by a high-accuracy stepwise numerical integration technique.   In this work, in order to generate membrane potential vs. time curves, an analytical solution similar to eqn. [9.32] was used, by modifying the boundary condition and averaging parameters, e.g., g_K and g_{Na}, over each time step.

$$\lozenge\,[t + \Delta t] = \lozenge\,[t]\ \exp\ [-\,\Delta t/\tau_m\,] \qquad\qquad [9.33]$$

9.15 PARTIAL REACTION RATE CONTROL

To proceed, one needs to investigate whether the chemical reactions can exert partial rate control.   For this purpose, it is necessary to analyze the coupling of the diffusion results with the receptor reaction kinetics, as shown below.

The ACh receptor is modeled as an allosteric protein with two sets of sites: one which corresponds to a conformation in a rest condition with a fractional penetrant saturation (Y_R) and another (Y_A) in which ACh is bound in an active (i.e., open channel) conformation (Reynolds and Karlin, 1978; Soddu and Vieth, 1980).   For an allosteric protein having two oligomers,

$$R \Leftrightarrow A$$

$$2R_0 + S \Leftrightarrow R_1 \qquad\qquad 2A_0 + S \Leftrightarrow A_1$$

$$R_1 + S \Leftrightarrow 2R_2 \qquad\qquad A_1 + S \Leftrightarrow 2A_2$$

$$\frac{dA_1}{dt} = 2k_{A1}\,SA_0 - k'_{A1}\,A_1 - k_{A2}\,SA_1 + 2k'_{A2}\,A_2 \qquad [9.34]$$

$$\frac{dA_2}{dt} = k_{A2}\,A_1\,S - 2k'_{A2}\,A_2 \qquad\qquad [9.35]$$

$Y_M = S/S_0$; $S = S_0 Y_M$;   *(Y_M (t) at the downstream location (X=1.0), is obtained from diffusion results; i.e., eqns. (9.19) and (9.20)).*
Boundary conditions: $A_1(0)=A_2(0)=0$. Let $y_1(t)=A_1/A_0$; $y_2 = A_2(t)/A_0$.

$$\frac{dy_1}{dt} = [2k_{A1}\,S_0\,Y_M] - [k'_{A1}]\,y_1 - [k_{A2}\,S_0\,Y_M]\,y_1 + [2k'_{A2}]\,y_2 \qquad [9.36]$$

$$\frac{dy_2}{dt} = [k_{A2}\,S_0\,Y_M]\,y_1 - [2k'_{A2}]\,y_2 \qquad [9.37]$$

At chemical equilibrium,

$$\frac{A_1\,[e]}{A_0} = \frac{2k_{A1}\,S_0\,Y_M}{k'_{A1}} \qquad [9.38]$$

$$\frac{A_2\,[e]}{A_0} = \frac{[k_{A2}]\,[S_0\,Y_M]\,[2k_{A1}\,S_0\,Y_M]}{[k'_{A1}]\,[2k'_{A2}]} \qquad [9.39]$$

Now, writing equations for the time-dependent fractional saturation of singly- and doubly-bound receptors, we obtain:

$$Y_{A1}\,[t] = \frac{A_1\,[t]\,/2A_0}{\dfrac{1}{\mathscr{L}}\,[1+ S/\,K_R\,]^2 + \dfrac{A_1\,[t]}{A_0} + \dfrac{A_2\,[t]}{A_0} + 1} \qquad [9.40]$$

$$Y_{A2}\,[t] = \frac{A_2\,[t]\,/A_0}{\dfrac{1}{\mathscr{L}}\,[1+ S/\,K_R\,]^2 + \dfrac{A_1\,[t]}{A_0} + \dfrac{A_2\,[t]}{A_0} + 1} \qquad [9.41]$$

where $A_0\,/\,R_0 = \mathscr{L} = 1.0$ and $K_R = 0.1\,S_0$ in this work.

This leads to a realization that the sodium and potassium conductivities should be individually formulated on the basis of the two classifications for open channels, corresponding to the singly- and doubly-bound receptors, respectively, i.e.,

$$g_{Na} = \bar{g}_{Na}\,Y_{A1}\,[t] \qquad [9.42]$$

$$g_K = \bar{g}_K\,Y_{A2}\,[t] \qquad [9.43]$$

That is, *a time separation of the conductivity profiles*, as is needed for epp simulation, can be effected on the basis of the probabilities of occurrence of singly- and doubly-bound species (Y_{A1} [t], Y_{A2} [t]) where $\overline{9}_{Na}$, 9_K are the saturation level conductivities for sodium and potassium, respectively. (All conductivity values are expressed relative to rest levels as baselines.) Equation [9.42] expresses the probability that a sodium channel is open, which is just equal to the probability that an ACh is bound locally by the receptor in the active conformation.

Under the nonequilibrium assumption and using the above equations [9.19, 9.20, 9.42, 9.43, 9.33], results are obtained by using the data given in Tables 9.4 and 9.5. (The time period under study was split into 500 separate intervals for computational purposes, using a 4th order Runge-Kutta integration routine (Vieth and Chotani, 1983). The results are shown in Fig. 9.13. The sodium conductivity leads and the potassium conductivity follows (the potential), as required. In contrast, when reaction equilibrium was assumed *a priori*, an unsatisfactory result was obtained; i.e., simulation research revealed that the conductivities could not be related to the potential in this simple manner (see Fig. 9.14).

Table 9.4   Reaction Parameters

$A = k'_{A1} = 7.5 \times 10^3$ (Wathey et al., 1979)

$B = k_{A2} S_0 Y_M = 3 \times 10^4 Y_M$ (Wathey et al., 1979)

$C = 2k'_{A2} = 10^3$ (Wathey et al., 1979)

$D = 2k_{A1} S_0 Y_M = 3 \times 10^8 \times 10^{-3} Y_M = 3 \times 10^5 Y_M$ (Wathey et al.,1979)

$E = A + B + C = 8.5 \times 10^3 + 3 \times 10^4 Y_M$

$F = AC = 7.5 \times 10^6$

$G = BD = 3 \times 10^4 \times 3 \times 10^5 \times Y_M^2 = 9 \times 10^9 Y_M^2$

Two other significant findings emerge: the value of D_{eff}/L^2 must be increased to a rather narrowly constrained limit and the conductivity relations become partially reaction rate dependent, thereby providing a means of model discrimination relative to the equilibrium assumption. Parameters such as D_{eff}/L^2, $9_{Na}, 9_K$, can be chosen in order to closely fit the standard results. In Fig. 9.13, the epp simulation results are shown for the finely adjusted values of D_{eff}/L^2, 9_{Na}, 9_K, equal to 0.84, 2.0, 2.0, respectively. Note that the estimated value of D_{eff} would be 8.4×10^{-8} cm^2/sec.

Table 9.5   EPP Constants

Constant	Symbol	Value chosen
peak amplitude above rest(mV)	$V_{max} - V_r$	20 (Changeux et al., 1983)
saturation conductivities:		
sodium (mmho/cm2)	$\bar{g}_{Na}$	2.0 (Vieth & Chotani,1983)
potassium (mmho/cm2)	$\bar{g}_K$	2.0 (Vieth & Chotani,1983)
membrane capacitance (μF/cm2)	C_M	1.0 (Hodgekin & Huxley,1952)
mean decay constant,start of falling phase (msec)	$C_M/(g_{Na}+g_K)$	ca.3.0 (Changeux et al., 1983)
sodium equilibrium potential(mV)	$V_r - V_{Na}$	-139 (Takeuchi, 1963)
potassium equilibrium potential (mV)	$V_r - V_K$	+10 (Takeuchi, 1963)
time axis shift factor (msec-1)	D_{eff}/L^2	0.84 (Vieth & Chotani,1983)
dimensionless rise time	θ_R	0.84 (Vieth & Chotani,1983)
dimensionless transmitter release time	θ_c	0.075 (Vieth & Chotani,1983)
dimensionless enzyme reactivity	μ	1.0 (Rosenberry,1979; Vieth & Chotani,1983)

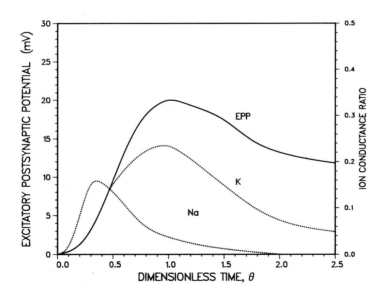

Figure 9.13   EPSP stimulation; non-equilibrium basis.

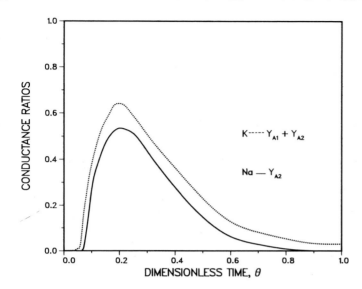

Figure 9.14   EPSP simulation; equilibrium basis.

It may be compared with the value 8.3 x 10^{-8} cm^2/sec estimated earlier in this chapter, page 314.   Given some uncertainty in overall thickness of the laminate structure due to junctional folds, it is probably too much to hope that a transport mean value of 1000Å (Lester, 1977) would represent the situation with such exactitude.   And the receptor binding reactions are not uniformly at equilibrium, so the full retardation effect on the diffusivity should perhaps not be applying; i.e., the fraction of the penetrant population which is immobilized will be somewhat smaller than its equilibrium value, so D_{eff} should be a bit larger than its equilibrium value.   However, Tshudy and von Frankenberg (1973) showed that, if the equilibrium distribution coefficient (k_D) is large enough, the membrane time lag tends to its maximum value, independent of the ratio of forward and reverse rate constants for the immobilization reaction.   In any event, the type of agreement found is clearly not just an accident, so perhaps another effect has come into play in the estimation.   For instance, our estimate of the equilibrium distribution coefficient for the receptor layer (i.e., 10^3) is possibly a bit low, due to incomplete purification of the receptor prior to reconstitution.   If so, the estimated equilibrium value of D_{eff} would be correspondingly reduced and the value found in simulation would exceed it a bit, as might be expected.

Independent of this issue, the effects of the reactions were explored thoroughly at the position X = 1.0. Detailed consideration of the conductivity relations vis-`a-vis the equilibrium and nonequilibrium assumptions reveals that the former is an oversimplification and that chemical gating steps must be included. Interestingly, if the constants in Table 9.4 could be altered at will, complex roots leading to oscillatory behavior could be determined.

9.16 EXTENSION TO PULSE CURRENTS

Mathematically, for a train of minipulses, the Duhamel Theorem can be applied to the transient solution for a unit step input, i.e., eqn. [9.19]. The ultimate solution takes the form,

$$\text{response} = \sum_{j=1}^{n} B_j \, \Delta f_{n+1-j} \qquad [9.44]$$

where $\Delta f_j = Y_M[1, j\,\Delta\Theta_c] - Y_M[1, \overline{j-1}\,\Delta\Theta_c]$,
Y_M = obtained from unit step response,
B_j = amplitude of jth pulse B_j = 1.0 firing ,   [9.45]
                                              = 0.0 rest,
n = total number of intervals in a train,
$\Delta\Theta_c$ = time span of each interval [= Θ_c/n].
For example, if n = 10, Θ_c = 1.0, $\Delta\Theta_c$ = 0.1, then the pulse train contains five pulse firings and five rests. For the various profiles shown in Fig. 9.15, each pulse is of unit amplitude. For reasons of simplicity, firing and rest state time intervals are assumed to be equal. When the time interval between pulses approaches zero or pulses overlap, the response will be the same as if a single unit pulse of Θ_c were the input. With some modifications in eqns. [9.44] and [9.45], response curves can be obtained with unequal time intervals of firing and rest states of vesicles. In this way, the continuous process model can be broken down into the very small intervals appropriate to multiple mini endplate currents.

Generally speaking, as soon as a pulse of S (i.e., ACh) sets in at the pre-synapse, it spreads diffusively through the cleft into the receptor sites at the post-synapse. As a result, active conformations come forth, ion channels open and the density of vacant receptor sites diminishes. This gives rise to the increase of conductances g_{Na}, g_K. Similarly, reverse phenomena take place as soon as the pulse is terminated upstream. The time evolution of the sodium and potassium conductances provides the key components in the relation of endplate current and endplate potential, as before.

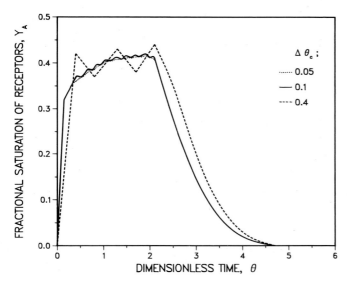

Figure 9.15 Response to a train of minipulses.

9.17 SIMPLE ANALYSIS OF OSCILLATORY BEHAVIOR

To begin the analysis, probabilities of occurrence of bound receptors are formulated as below:

$$\frac{dy_1}{dt} = D - \alpha y_1 - By_1 + Cy_2 \qquad \frac{dy_2}{dt} = By_1 - Cy_2 \qquad y_1 = \frac{1}{B}[dy_2/dt + Cy_2] \qquad [9.46]$$

$$\frac{d^2 y_2}{dt^2} = B\frac{dy_1}{dt} - C\frac{dy_2}{dt} = B[D - \frac{\alpha + B}{B}[dy_2/dt] + Cy_2] + Cy_2] - Cdy_2/dt$$

$$= BD - [\alpha + B + C]dy_2/dt - [\alpha C]y_2 \qquad [9.47]$$

where $D = 2k_{A1} S_0 Y_M$, $\alpha = k'_{A1}$, $B = k_{A2} S_0 Y_M$ and $C = 2k'_{A2}$, or:

$$\frac{d^2y_2}{dt^2} + E \frac{dy_2}{dt} + Fy_2 = G \qquad [9.48]$$

where $E = \alpha + B + C$, $F = \alpha C$, $G = BD$.

Over a small interval of time, the solutions are:

$$y_2 = \frac{G}{F} + C_1 \exp[m_1 t] + C_2 \exp[m_2 t] \qquad [9.49]$$

where:

$$m_{12} = \frac{-E \pm [E^2 - 4F]^{1/2}}{2} \qquad [9.50]$$

and, when $E^2 - 4F > 0$, complex roots appear which can lead to oscillatory behavior. (It seems possible that neurotoxins could bring about parametric changes which would produce these results.) To proceed with the solution,

$$t = 0, \quad y_2 = 0, \quad y_1 = 0, \quad \frac{dy_2}{dt} = 0 \qquad [9.51]$$

$$\frac{G}{F} = -[C_1 + C_2] \qquad [9.52]$$

$$m_1 C_1 + m_2 C_2 = 0 \qquad C_2 = -\frac{m_1}{m_2} C_1 \qquad [9.53]$$

$$C_2 = -\frac{m_1}{m_2} \frac{G}{F} \frac{m_2}{m_1 - m_2} \qquad [9.54]$$

$$y_1 = \frac{1}{B} \{ C_1 m_1 \exp[m_1 t] + C_2 m_2 \exp[m_2 t] + Cy_2 \}$$

$$y_2[\Theta] = \frac{G}{F} + C_1 \exp[\hat{m}_1 \Theta] + C_2 \exp[\hat{m}_2 \Theta] \qquad [9.55]$$

$$y_1[\Theta] = \frac{1}{B}\{C_1 m_1 \exp[\hat{m}_1 \Theta] + C_2 m_2 \exp[\hat{m}_2 \Theta] + Cy_2\} \qquad [9.56]$$

Equations [9.56] and [9.57] will display unstable oscillatory behavior which would correspond to spasmodic response and progressive loss of muscle function at the affected junctions.   This is precisely what can be experimentally observed in nerve excitation, involving *muscarinic* cholinergic receptors (Monnier, 1984).

Simple electrical circuitry shows that an electrical cathode on a nerve elicits in the latter an "excitation process," adequately expressed by a double exponential (as in the graph of Fig. 9.13).   But under special conditions, such as low calcium in the surrounding fluid (alluding to its rather central role), the excitation process is characterized by a damped sine wave (Monnier, 1984).

If the damping is further reduced, permanent electrical oscillatory behavior appears which then may lead to the periodic firing of propagated impulses along the nerve fibre.   When the membrane potential of a fiber, or a cell, is reduced by the cathode of an external circuit or by an activating channel mediator, the damping factor is reduced and periodic excitation follows.   When the membrane potential is increased, whether by an anode or by an inhibitory mediator, the damping factor is increased and excitation is not forthcoming.   In other words:

Reduced membrane potential = reduced damping factor and
   activation.
Increased membrane potential = increased damping factor and
   inhibition.

At this juncture, it is time to leave the nicotinic and muscarinic receptors per se and to go on to a consideration of β-adrenergic receptor processes, with special emphasis on the mammalian myocardium.   These processes include a muscarinic cholinergic receptor component, so the modeling undertaken so far may ultimately be useful there too.

9.18 REFERENCES

Adams, P.R., H.M. Pinsky and W.D. Willis, Jr., (eds.) "Information Processing the Nervous System," Raven Press: New York, p. 108 (1980).

Anholt, R., J. Lindstrom and M. Montal, *Eur. J. Biochem.*, **109**, 481 (1980).

Barman, T.E., "Enzyme Handbook, II," Springer: New York, p.510 (1969).

Barrie, J.A., J.D. Levine, A.S. Michaels and P. Wong, *Trans. Faraday Soc.*, **59**, 869 (1963).

Bhatia, D. and W.R. Vieth, *J. Mem. Sci.*, **6**, 351 (1980).

Carslaw, H.S. and J.C. Jaeger, "Conduction of Heat in Solids," 2nd ed., Oxford University Press: Clarendon (1959).

Changeux, J.P., F. Bon, J. Cartaud, A. Devilliers-Thiery, J. Giroudat, T. Heidmann, B. Holton, H.O. Nghiem, J.L. Popat, R. VanRapenbusch and S. Tzartos, *Symposia on Quant. Biol.* C.S.H., **48**, 35 (1983).

Couteaux, R., *Experimental Cell Research*, **Suppl. 5**, 323 (1958).

Eccles, J.C. and J.C. Jaeger, *Proc. R. Soc.*, **B148**, 38 (1958).

Ehrenstein, G., *Physics Today*, **October**, 33 (1976).

Elliot, J., S.G. Blanchard, W. Wu, J. Miller, C.D. Straden, P. Hartig, H.-P. Moore, J. Racs and M.A. Raftery, *Biochem. J.*, **185**, 667 (1980).

Grant, R.A., R.W. Cox and R.W. Horne, *J. Royal Microscopical Soc.*, **87**, Part 1, 143 (1967).

Heidmann, T. and J.-P. Changeux, *Ann. Rev. Biochem.*, **47**, 317 (1978).

Hirose, S., W.R. Vieth and M. Takao, *J. Molec. Catal.*, **18**, 11 (1983).

Hirose, S. and W.R. Vieth, in *Appl. Biochemistry and Biotechnology*, **9**, 81 (1984).

Hodgkin, A.L. and A.F. Huxley, *J. Physiol. (London)*, **117**, 50 (1952).

Katz, B., "Nerve, Muscle and Synapse," McGraw-Hill: New York, p. 140 (1966).

Kuffler, S.W. and D. Yoshikami, *J. Physiol.*, **251**, 465 (1975).

Lester, H.A., *Sci. American*, **236**, 106 (1977).

Lindstrom, J., R. Anholt, B. Einarson, A. Engel, M. Osame and M. Montal, *J. Biol. Chem.*, **255**, 8340 (1980).

Ludolph, R.A., W.R. Vieth, K. Venkatasubramanian and A. Constantinides, *J. Mol. Catal.*, **5**, 197 (1979).

Monnier, A., Personal Communication (1984).

O'Brien, R.D. (ed.), "The Receptors," Plenum Press: New York (1980).

Reynolds, J. and A. Karlin, *Biochemistry*, **17**, 2035 (1978).

Rosenberry, T.L., *Biophys. J.*, **26**, 263 (1979).

Soddu, A. and W.R. Vieth, *J. Mol. Catal.*, **7**, 491 (1980).

Takeuchi, N., *J. Physiol. (London)*, **167**, 141 (1963).

Tshudy, J.A. and C. von Frankenberg, *J. Polym. Sci., Polym. Let. Ed.*, **23** 525 (1985).

Vieth, W.R. and G.K. Chotani, *J. Mol. Catalysis*, **19**, 171 (1983).

Vieth, W.R. and G.K. Chotani, *J. Mol. Catalysis*, **22**, 27 (1983).

Wathey, J.C., M.N. Nass and H.A. Lester, *Biophys. J.*, **27**, 145 (1979).

Weadock, K., PhD. Thesis in Biomedical Engineering, Rutgers U. (1986).

Weadock, K., F.H. Silver and D. Wolff, *Biomaterials*, in press (1987).

Light in the Spirit, light in the dawning.
Light in creation of the new morn.

Trad. Ballad

10

MAMMALIAN EXCITABLE MEMBRANES

10.0 INTRODUCTION

Excitable tissues such as the myocardial sino-atrial node and Purkinje fiber possess special membranes that enable them to maintain very large calcium concentration gradients. The calcium level inside a cell during a depolarization-repolarization cycle or a contraction-relaxation cycle fluctuates approximately between 0.1 and 10 microM. In Purkinje fiber, these changes in free cytoplasmic calcium levels are generally attributed to calcium influx and efflux through electrically and metabolically controlled channels in the sarcolemma as well as to a calcium release and uptake mechanism in the sarcoplasmic reticulum. The biochemical control mechanisms involved in these calcium movements are not yet understood in ultimate detail. However, two of the principal regulatory control agents in these membranes (cAMP-dependent protein kinase and calmodulin) have been identified (Watanabe et al., 1981; Williams, 1986; Michiel and Wang, 1986). Both calcium uptake and calcium-stimulated ATPase activities in the sarcoplasmic reticulum are markedly increased by protein kinase and calmodulin. Sarcolemmal calcium-pump activities are also stimulated by these two substances. These observations illustrate the type of metabolic controls regulating membrane calcium transport.

In considering the several myocardial calcium currents it is important to distinguish their locales of origin. The primary pacemaker potential is generated at the sino-atrial node which communicates with the Purkinje fiber (where strong contractions occur) by passing along the electrical signal through the atrioventricular, or AV node, the conduction and refractory characteristics of which slow the spread of the impulse and thus prevent inordinately rapid ventricular rates. It has been established that similar hormonal stimulation effects take place at both the SA node and the Purkinje fiber; the modulating influence of the cyclic nucleotide system, through its interaction with calcium currents, is known in detail for the latter. It seems probable that the same, or a very similar, biochemical control system functions at both locales.

10.1 ION CHANNELS AND AUTONOMIC RECEPTORS OF CARDIAC PURKINJE FIBER

Together with a gated calcium channel which is transmitter-dependent, the ionic channels that have been found in Purkinje fiber are the voltage-dependent sodium channel, voltage-dependent potassium channel and a membrane potential-dependent calcium channel (Hagiwara and Ohmori, 1982; Tillotson, 1979; Cachelin et al., 1983; Hagiwara and Byerly, 1981; Tsien, 1983). Also located on the membrane are two electrogenic ion exchange devices. Sodium-potassium-ATPase works as a sodium-potassium pump that transports three sodium ions outward in exchange for two potassium ions transported inward. A phosphorylated intermediate of the enzyme sodium-potassium-ATPase occurs in the transport cycle, its phosphorylation being sodium-dependent and its dephosphorylation being potassium-dependent (Sperelakis, 1979). The calcium-sodium mechanism exchanges one internal calcium ion for 2, 3 or 4 external sodium ions via a membrane carrier molecule. This reaction is facilitated by ATP, but ATP is not hydrolyzed. Instead, the energy for the pumping of calcium against its large electrochemical gradient comes from the sodium electrochemical gradient. That is, the uphill transport of calcium is coupled to the downhill movement of sodium. Effectively, the energy required for this calcium ion movement is derived from the sodium-potassium-ATPase. Thus, the sodium-potassium pump which uses ATP to maintain the sodium electrochemical gradient, indirectly helps to maintain the calcium concentration gradient.

It has been discovered that catecholamine bound to the receptor activates the enzyme adenylate cyclase (Murad et al., 1962) which produces cAMP, and cAMP then activates the protein kinase which phosphorylates the gating proteins of the chemically-gated calcium channel (e.g., Ikemoto and Goto, 1978; Tsien, 1973; Tsien and Weingart, 1976; Yamasaki et al., 1974; Vogel and Sperelakis, 1981; Trautwein et al., 1982; Reuter, 1983; Curtis and Catterall, 1985). Phosphodiesterase inactivates cAMP by converting it to 5'-AMP and PDE is activated by intracellular free calcium ions via a mechanism that involves an intervention of calcium-sensitive calmodulin. Figure 10.1 shows the reactions described above.

It is well known that the Purkinje fiber of the mammalian heart possesses pacemaker activity (DiFrancesco, 1981) as well as its signal conductive property. The genesis of automaticity in Purkinje fibers is somewhat different from that in nodal cells. Normally the automaticity of cells lower in hierarchy (e.g., Purkinje cells, AV nodal cells) is latent because the cells are driven at higher rates by the primary pacemaker (SA node). Very shortly, the issue of biochemical control of pacemaker potential in the SA node is explored. But first, a few more words on biochemical control of receptor-operated calcium channels (ROCs) in the Purkinje fiber is in order.

Cyclic AMP is known to activate cAMP-dependent protein kinase (A-PK) which consequently phosphorylates catalytic subunits of ROCs, leading to the opening of ROCs. In Purkinje fiber, the elevation of intracellular cAMP concentration results in a prolonged and heightened plateau phase of action potential (Nargeot et al.,1983; Cachelin et al., 1983; Brum et al., 1984), as well as stronger contraction. This leads to the notion that cAMP and A-PK may play even more significant roles for the SA node since inward calcium current flowing through ROCs whose activation requires the presence of cAMP and A-PK could constitute the basis for the intrinsic oscillation of membrane potential.

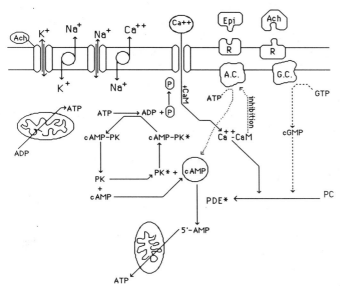

Figure 10.1   Biochemical control mechanism for receptor-operated channels (ROCs) at the Purkinje fiber.

10.2 ION CHANNELS AND AUTONOMIC RECEPTORS OF SA NODE

Sharing the same functions of ionic channels and receptors with other parts of the myocardia, the SA node's spontaneous generation of electrical activity may be explained in biochemical terms at the subcellular level in relation to the cyclic nucleotide system. This becomes readily apparent when hormonal effects are considered. For instance, the effect of epinephrine on cardiac response is an increase in the calcium current, which increases internal calcium, amounting to increase in the inward current during diastole which leads to faster diastolic depolarization. Thus, one explanation is that myocardial receptor-operated channels, or ROCs, are readily modulated by hormones, in that channel opening is facilitated through the intervention of intracellular cAMP level which is, in turn, determined by adenylate cyclase activity triggered by β-agonists.

Knowing that calcium current at the SA node is modulated by neurohormones, it has been suggested that the principal action of epinephrine on slow inward current is to increase the maximal conductance of the calcium channel. This can occur either by increasing the maximal conductance of each slow inward current channel or by creating more channels, each with a constant single-channel conductance (Giles and Noble, 1976; Erik et al., 1976; Kass and Wiegers, 1982; Bean et al., 1984; Kokubun and Reuter, 1984). (Potassium channels are presumably always open as a background leak current, but regulated by transmitters.)

The action potential of the SA node does not show a rapid upstroke which is a result of a fast inward sodium current, as observed in Purkinje fiber and ventricular cells. This shows that the sodium current in the SA node does not contribute to the construction of action potential as much as it does in other cardiac cells. AV nodal cells, which are also capable of producing pacemaker potential, are known to have calcium channels through which both sodium and calcium flow (Sperelakis, 1984). Therefore, the membrane permeability of the SA node to sodium ions is much smaller than that of Purkinje fiber, with sodium channels being not so specific for sodium ions. On the other hand, the calcium conductivity is high enough to play an important role in generating an action potential. Calcium current is strongly affected by intracellular cAMP, and the calcium channel is believed to be opened by a protein kinase which is activated by cAMP. Cyclic AMP consequently causes an influx of calcium ions which then regulates the action of cAMP by activating phosphodiesterase, a mechanism that involves a regulatory protein, calmodulin. This interaction sets up a self-oscillatory feedback system between cAMP and free intracellular calcium ions that makes possible a genesis of pacemaker potential.   This is graphically shown in Fig. 10.2.

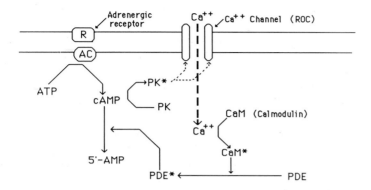

Figure 10.2   Simplified biochemical mechanism for control of pacemaker potential.

We are getting a bit ahead of ourselves, so let us retrace our steps.   Theories of what constitutes diastolic depolarization at the SA node are diverse.   By analogy with the Purkinje fiber case, spontaneous generation of depolarization had been considered to arise from the particular characteristics of myocardial membrane permeability to potassium and sodium ions.   Diastolic depolarization had sometimes been attributed to sodium leakage current because of the belief that, in all cardiac cells, there is an inward leakage current, which is presumably carried by positively charged sodium ions.   If this leakage current is not offset in diastole by an electrically equal outward ion flux, then diastolic depolarization will ensue (Wallace, 1982).   But, unlike Purkinje fiber where sodium current is the dominant factor for construction of the action potential, SA nodal cells show a slow upstroke of action potential mainly consisting of strong inward calcium current while concurrent sodium flow is considered a minor fraction of the total current that is carried by both ions (Vassalle, 1977).   On the other hand, a sudden decrease in potassium conductance immediately after repolarization, which is another prominent feature of Purkinje fiber, has also been considered to be responsible for diastolic depolarization at the SA node, due to some of the electrophysiological similarities of the latter to Purkinje fiber;   however, it has been determined that such a drop in potassium conductance is so slight as to be almost undetectable in the SA node (Yanagihara et al., 1980).

In contrast, a gradual increase of slow inward calcium current, i_s, during diastole prior to the upstroke was strongly suggested as a candidate for the triggering factor of pacemaker depolarization (Yanagihara et al., 1980). This suggestion is supported by the findings that, during diastole, i_s shows gradual increase whereas contemporary decrease in potassium conductance is hardly observed. Thus attention has come to the fact that the dominant current of the SA node is carried by calcium ions and is facilitated by unique functions of calcium channels that are activated by both electrically responsive (systole) and biochemically responsive (diastole) influences. Therefore it becomes clear that the myocardium possesses both potential-dependent calcium channels (PDCs) and receptor-operated calcium channels (ROCs) whose roles are additive in some manner.

The suggestion that gradual increase of i_s during diastole is the main cause for diastolic depolarization gives rise to an idea that both diastolic depolarization and systolic depolarization are constructed by inward calcium current, and that the two stages of depolarization take place through successive *dual* mechanisms because systolic depolarization only occurs at the potential of -40 mV or above and diastolic depolarization occurs between -60 mV and -40mV. Moreover, voltage clamp experiments have shown a slow inward current, which is a potential-dependent calcium current, being recorded on depolarizations from -40 mV (Brown et al., 1979; Yanagihara et al., 1980). Thus, one can envision that systolic depolarization is attributable to the calcium current flowing through PDCs and diastolic depolarization is attributable to a precursor calcium current which is facilitated by a mechanism that is biochemically controlled. Biochemical control of diastolic depolarization is then introduced by incorporating the role of ROCs as a passage for the initial calcium influx that can raise the membrane potential from   -60 mV to -40 mV. The role of ROCs at this early stage of depolarization is emphasized since it is also useful for explaining automaticity, which can be considered as a continuous intrinsic oscillation of membrane potential. Thus, ROCs provide channels for calcium influx which generates a certain degree of depolarization that is enough for activation of PDCs through which additional calcium current flows to complete an action potential (Nho, 1987).

In this manner, ROCs are considered responsible for initiation of diastolic depolarization, while PDCs, on the other hand, are responsible for the upstroke and height of action potential. Therefore, automaticity may be attributable to a mechanism that powers and controls the repetitive process of opening and closing ROCs. As evidence for the validity of this postulate, β-agonists are known to induce a positive chronotrophic effect, by a mechanism that utilizes ROCs in interpreting their messages and responding to them. The degree of chronotrophy is determined by a factor that eventually affects the slope of diastolic depolarization, indicating a heightened role of ROCs in that phase.

A study performed on rabbit SA node shows persistent rhythmic oscillations of action potential in normal Tyrode solution (2.7 mM K$^+$), turning into reduced local oscillations in K$^+$-rich Tyrode solution (23 mM K$^+$) (Fleckenstein, 1985). Increased extracellular potassium concentration definitely raises resting and threshold potential so that PDCs become less likely to be activated. Consequently, at higher resting potential, the generation of propagated pacemaker action potential is no longer possible. Instead, permanent local subthreshold oscillations demonstrate the persistence of intrinsic nodal automaticity. The fact that continuous subthreshold oscillations persist without concomitant generation of action potential lends support to the view that ROCs are solely responsible for the subthreshold oscillation and that there exists a separate mechanism that produces such oscillation.

10.3 β-ADRENERGIC INFLUENCE ON NOMOTOPIC AUTOMATICITY

Diastolic depolarization is accelerated when calcium influx is increased by norepinephrine whose activity is to stimulate membrane-bound adenylate cyclase to generate cAMP which, in turn, catalyzes the opening process of receptor-operated calcium channels. Likewise, epinephrine tends to increase the slope of the diastolic depolarization phase. Therefore, the existence of hormonally-controlled ROCs at the SA nodal membrane is clear. The fact that there exists a biochemical mechanism that carries out the positive chronotropic effect implies that the same biochemical mechanism works for the ongoing intrinsic automaticity, so that the strengthened hormonal effect must come from a magnification of the already-existing mechanism which prefigures it. That, in short, is the rationale for the conceptualization so far. To restate it briefly: It is apparent that the myocardium possesses both potential-dependent calcium channels (PDCs) and receptor-operated calcium channels (ROCs) and that their roles are additive. ROCs provide channels for calcium influx which generates a certain degree of depolarization that is sufficient for the activation of PDCs through which additional calcium current flows to complete an action potential.

Myocardial ROCs are readily modulated by hormones in that channel opening is affected by intracellular cAMP level which is determined by adenylate cyclase activity triggered by β-agonists. Cyclic AMP activates cAMP-dependent protein kinase (A-PK) which phosphorylates the catalytic subunits of ROCs which then leads to the opening of calcium channels for the consequent inward flow of calcium current. A balancing outward calcium flow occurs in a later phase via active transport of a calcium-calmodulin complex.

The interaction of calmodulin with its target proteins requires the integrity of different portions of the calmodulin molecule. Calmodulin-regulated enzymes can be divided into three classes according to their abilities to bind with and to be activated by calmodulin fragments: (1) enzymes which are activated by the C-terminal fragment, such as the Ca^{++}-ATPase and phosphorylase kinase; (2) enzymes which require both halves of the molecule, such as cAMP phosphodiesterase and myosin light chain kinase; and (3) enzymes whose interaction with calmodulin fragments is too weak to be detected by activation, such as calcineurin and the multiprotein kinase. Thus, different enzymes may be activated by different calmodulin conformers and the stepwise changes exhibited by calmodulin at different calcium levels can be used to regulate different metabolic pathways (Ni and Klee, 1985).

We must leave this topic now, to go on to a consideration of photoreceptors, where calcium again is found to play a central role. The discussion continues, in the following section.

10.4 PHOTORECEPTORS

Advances in elucidating the molecular basis of visual transduction have come from two sources; studies of bacteriorhodopsin and of rhodopsin in the vertebrate retina. Bacteriorhodopsin is contained in the purple membrane, a specialized patch in the membrane of Halobacteria. A protein of 248 amino acids, it contains a light-absorbing chromophore called retinal, which is identical to that found in the rod photoreceptor cells of the vertebrate eye.

A single photon of light excites the chromophore, causing it to change in conformation; in so doing the chromophore transfers two hydrogen ions out of the cell (Barnard et al., 1983). The question is still unanswered as to what path the ion takes through the membrane in moving from the inside to the outside of the cell. This conformational change in a single photoreceptor protein molecule leads to the transient closure of many sodium channels in the plasma membrane of the rod outer segment (Hubbell and Brownds, 1979; Fein and Szuts, 1982).

In rod photoreceptor cells of vertebrates, the rhodopsin is not located in the external membrane but in the membrane of the disk, an intracellular organelle.   Nonetheless, photoisomerization of the 11-cis retinal chromophore of rhodopsin to the all-trans form gives rise to a potential change (a hyperpolarization) in the external membrane of the photoreceptor neuron that is essential for signaling.   The resulting hyperpolarization of plasma membrane is conveyed to the synapse at the outer end of the rod and communicated to other cells of the retina.   To accomplish this, a chemical messenger - a transmitter or an ion - may carry information from the disk membrane to the external membrane. Evidence points strongly to the participation of two messengers; i.e., calcium and cGMP.

It has been proposed that calcium ions taken up by disks in the dark serve as the transmitter.   According to this model (Yoshikami et al., 1980), light leads to the release of many calcium ions into the cytosol, which then diffuse to the plasma membrane and block sodium channels.   However, disks have not been shown to sequester calcium in the dark. Overall findings point to an important role of calcium in visual excitation but raise the question as to whether calcium acts alone in conveying the signal to the plasma membrane.

Cyclic GMP came into prominence for sharing the role with calcium as a transmitter because of a number of biochemical and electrophysiological findings.   Firstly, the rod outer segments (ROS) contain an unusually high concentration ($70\mu M$) of cGMP.   Secondly, the cGMP level in ROS decreases markedly on illumination (Woodruff and Brownds, 1979).   Thirdly, ROS possess cGMP phosphodiesterase.

Light activates a phosphodiesterase (PDE) in ROS, which hydrolyses cGMP.   This is thought to have a role in closing the sodium channels and hyperpolarizing the cell.   A single photo-excited rhodopsin molecule activates several hundred molecules of PDE.   The photoactivation of PDE is mediated by transducin, a peripheral membrane protein whose activation requires the exchange of GDP for GTP.   In its dependence on GTP and its sensitivity to inhibition by cholera toxin, transducin resembles the G protein of the adenylate cyclase system.   This similarity suggests that the transducin and the adenylate cyclase system of other tissues belong to the same family of signal-coupling proteins and that the activation of cGMP PDE by   light resembles the activation of adenylate cyclase by hormones or transmitters.

Fig. 10.3 shows an interesting comparison of the activation of PDE (A) by light in retinal ROS with that of adenylate cyclase (B) by a hormone such as epinephrine (Stryer, 1983).

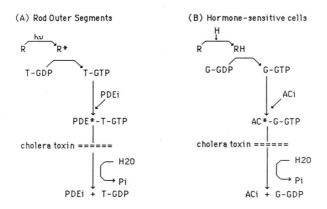

Figure 10.3   (R) Rhodopsin or hormone receptor;   (R*) photoexcited rhodopsin;   (RH) hormone-receptor complex;   (T) transducin;   (G) Stimulatory G protein;   (ACi and AC*) Inhibited and activated forms of AC;   (PDEi and PDE*) Inhibited and activated forms of PDE.

10.5 BRAIN RECEPTORS

The complexity of the nervous system emerges from many developmental steps that encompass determination, differentiation, formation of connections, cell death, synapse elimination, and fine-tuning of the remaining synapses.   These steps reflect lineage, early developmental history, later competitive interactions, and other epigenetic modifications.   The resulting heterogeneity of cellular properties in the brain is often referred to as neuronal diversity (Kandel, 1983).   Neurons differ in their signaling capabilities.   Some neurons generate action potentials; others do not.   All these differences can be traced to the family of specific ion channels present in the membrane of a given neuron, and neurons differ in their ion channels.

Neurons differ in the chemical transmitters that they synthesize. Some neurons synthesize one of a large family of small molecule transmitters; other neurons synthesize one and often several peptide transmitters; still others synthesize a combination of these transmitters. Neurons differ from one another in the connections they make with their target cells. Neurons differ in the receptors they have to small molecule transmitters, peptides, and hormones. Not only do cells respond differently to different transmitters, but they can respond differently to the same transmitter. Neurons differ in structure, in the size of the cell body, in the presence or absence of axons, and in the number and shape of dendrites. Finally, neurons are thought to have distinctive recognition molecules that distinguish neurons in the various regions of the brain from each other during development and that also distinguish the position of each cell or cell grouping within each region.

In addition to the diversity of brain neurons, the cells of the brain make extensive use of an unusual class of mRNA that is encountered only infrequently in other cells. Whereas in other tissues the mRNA that is translated into protein almost invariably is polyadenylated (poly[A]$^+$), the brain contains a large number of mRNAs lacking this poly[A]$^+$ tail.

In mammalian brain cells, nicotinic AChRs and GABA receptors are believed to play an important role in processing information even though this brain nicotinic AChR cannot be assumed to have a direct channel-gating property like other nicotinic AChRs. On the other hand, the GABA receptor of the vertebrate brain is known to open a chloride ion channel at a major class of GABA - mediated inhibitory synapses (Barnard et al., 1983).

The components responsible for the transduction or other transmembrane signaling by the receptor are, in most cases, another set of proteins working in tandem. Thus, many receptors employ cyclic nucleotide second messengers. The cyclic nucleotide system that causes the generation of cAMP and cGMP in brain neurons is illustrated in Fig 10.4 (Woody, 1982).

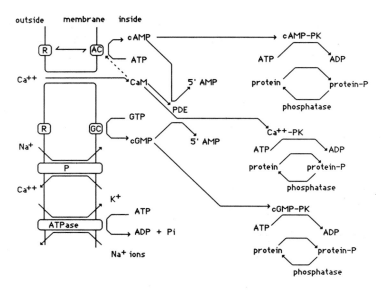

Figure 10.4   Calcium regulation of brain cAMP metabolism (hypothetical).

Action of enzymes in relation to cyclic nucleotide metabolism in the brain requires the mediation of calmodulin.   Calmodulin is a relatively small regulatory protein which mediates many calcium effects, including the stimulation of brain adenylate cyclase (AC) and calcium-sensitive phosphodiesterase.   In both cases, calmodulin forms a complex with the apoenzyme in the presence of calcium.   The calcium-sensitive PDE is a tetramer comprised of two catalytic and two calmodulin units.   The brain may contain distinctive forms of AC, one calmodulin-sensitive and another calmodulin-insensitive.   Neither form has been clearly characterized, and the relationship between the two forms remains to be established.

The stimulation of phosphodiesterase by calmodulin does not require any other cofactor.   In contrast, stimulation of AC by calmodulin appears to require a regulatory subunit in addition to the catalytic subunit.   Preliminary evidence suggests that this subunit is the guanyl nucleotide regulatory subunit.   There are a number of similarities between calcium and hormone stimulation of AC.   Both types of regulation are mediated by specific receptors that are distinct entities, and both apparently require a guanyl nucleotide regulatory subunit.

A hypothetical scheme depicting the regulation of brain cAMP metabolism by the cellular flux of calcium ions is shown in Fig. 10.5.

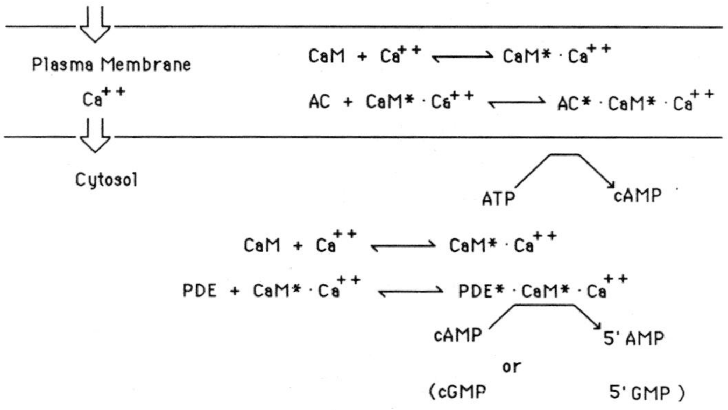

Plasma Membrane

Ca^{++}

$CaM + Ca^{++} \longleftrightarrow CaM* \cdot Ca^{++}$

$AC + CaM* \cdot Ca^{++} \longleftarrow AC* \cdot CaM* \cdot Ca^{++}$

Cytosol

ATP   cAMP

$CaM + Ca^{++} \longleftarrow CaM* \cdot Ca^{++}$

$PDE + CaM* \cdot Ca^{++} \longleftrightarrow PDE* \cdot CaM* \cdot Ca^{++}$

cAMP   5′ AMP

or

(cGMP   5′ GMP)

Figure 10.5 Role of calmodulin.

10.6 CONCLUDING REMARKS

It can be readily noticed that there are some similarities underlying the excitation mechanisms between photoreception and hormonal reception at nerve endings. It seems that retinal rod makes use of calcium and cGMP as second messengers, whereas calcium and cAMP play the messenger role for other excitable cells that have been mentioned. Light, in photoreception, triggers the release of calcium ions, and light activates a cGMP-PDE via a mechanism that requires the action of transducin. The calcium ions released then diffuse to the plasma membrane and block sodium channels to hyperpolarize the cell.

Since the question remains unanswered as to whether calcium acts alone or requires the involvement of cGMP to convey the signal to the plasma membrane, one might argue that the ROS contain a negative feedback system either between calcium and cGMP or between calcium, cGMP and light. The latter assumption might involve the action of light as a second messenger or could be based on the hypothesis that calcium demands the participation of cGMP for the transport of the signal to the plasma membrane.

The postsynaptic membrane of the brain neuron is thought to possess both the properties of adrenergic receptor and of muscarinic or nicotinic cholinergic receptor. This explains well how the brain neurons might work, as illustrated in Fig.10.4, in terms of the interaction between transmitters and the cyclic nucleotide system, in association with channel gating.

It has been discovered that peptides, ranging in length from 2 to 100 amino acids, also serve as chemical transmitters within the nervous system, and these peptides are likely to be very important for understanding brain function. A given peptide functions in three overlapping ways; firstly, as a neurotransmitter by acting over very short distances (300-500 Å) on neighboring nerve cells; secondly, as a local hormone by diffusion over somewhat larger distances (1-2mm); thirdly, as a neurohormone by being released into the blood stream to act on distant targets. Peptides are released in a calcium-dependent manner and they act upon specific receptors in the postsynaptic cell. The application of membrane diffusion- reaction principles to these phenomena is a challenge whose importance should be apparent. It is hoped that the reader will be stimulated to take up this task!

10.7 REFERENCES

Barnard, E.A., E. Beeson, G. Bilbe, D.A. Brown, A. Constanti, B.M. Conti-Tronconi, J.O. Dolly, S.M.J. Dunn, F. Mehraban, B.M. Richards and T.G. Smart, *Symposia on Quant. Biol.*, Cold Spring Harbor, **48**, 109 (1983).

Bean, B.P., M.C. Nowycky and R.W. Tsien, *Nature (Lond.)*, **307**, 371 (1984).

Brown, H.F., D. DiFrancesco and S.J. Noble, *Nature*, **280**, 235 (1979).

Brum, G., W. Osterrieder and W. Trautwein, *Pflugers Arch. Eur. J. Physiol.*, **401**, 111 (1984).

Cachelin, A.B., J.E. dePeyer, S. Kokubun and H. Reuter, *Nature*, **304**, 462 (1983).

Curtis, B.M. and W.A. Catterall, *Proc. Natl. Acad. Sci. USA*, **82**, 2528 (1985).

DiFrancesco, D. *J. Physiol.*, **314**, 359 (1981).

Erik, R.Ten, H. Nawrath and T.F. McDonald, *Pfluger Arch.*, **361**, 207 (1976).

Fein, A. and E.Z. Szuts, "Photoreceptors: Their Role in Vision," Cambridge Univ. Press (1982).

Fleckenstein, A., in *Calcium in Biological Systems*, Plenum Press, p. 459 (1985).

Giles, W. and S.J. Noble, *J. Physiol.*, **261**, 103 (1976).

Hagiwara, S. and L. Byerly, *Ann. Rev. Neurosci.*, **4**, 69 (1981).

Hagiwara, S. and H. Ohmori, *J. Physiol.* **331**, 231 (1982).

Hubbell, W.L. and M.D. Brownds, *Ann. Rev. Neurosci.*, **2**, 17 (1979).

Ikemoto, Y. and M. Goto, in *Recent Advances in Studies on Cardiac Structure and Metabolism*, **11**, 57 (1978).

Kandel, E.R., *Symposia on Quant. Biol.*-C.S.H., **48**, 891 (1983).

Kass, R.S. and S.E. Wiegers, *J. Physiol.*, **322**, 541 (1982).

Kokubun, S. and H. Reuter, *Proc. Natl. Acad. Sci. USA*, **81**, 4824 (1984).

Michiel, D.F. and J.H. Wang, *Intracellular Calcium Regulation*, Manchester University Press (1986).

Murad, F., Y.-M. Chi, T.W. Rall and E.W. Sutherland, *J. Biol. Chem.*, **237**, 1233 (1962).

Nargeot, J., J.M. Nerbonne, J. Engels and H.A. Lester, *Proc. Nat. Acad. Sci USA*, **80**, 2395 (1983).

Ni, W.C. and C.B. Klee, *J. Biol. Chem.*, **260**, 6974 (1985).

Nho, K., Ph.D. Thesis in Chemical and Biochemical Engineering, Rutgers U. (1987).

Reuter, H., *Nature (Lond.)*, **301**, 569 (1983).

Sperelakis, N., *The Cardiovascular System*, **1**, 187 (1979).

Sperelakis, N., *Physiol. and Pathophysiol. of Heart*, 106 (1984).

Stryer, L., *Symposia on Quant. Biol.*-C.S.H., **48**, 841 (1983).

Tillotson, D., *Proc. Natl. Acad. Sci. USA*, **77**, 1497 (1979).

Trautwein, W., J. Taniguchi and A. Noma, *Pfluegers Arch.*, **392**, 307 (1982).

Tsien, R.W., *Nature (London) New Biol.*, **245**, 120 (1973).

Tsien, R.W. and R. Weingart, *J. Physiol. Lond.*, **260**, 117 (1976).

Tsien, R.W., *Ann. Rev. Physiol.*, **45**, 341 (1983).

Vassalle, M., *Am. J. Physiol.*, **233(6)**, H625 (1977).

Vogel, S. and N. Sperelakis, *J. Mol. Cell. Cardiol.*, **13**, 51 (1981).

Wallace, A.G., "The Heart," 5th edn., McGraw-Hill, p. 120 (1982).

Watanabe, A.M., J.P. Lindemann, L.R. Jones, H.R. Besch, Jr. and J.C. Bailey, *Am. Physiol. Soc.*, **189** (1981).

Williams, R.J.P., *Calcium and the Cell*, John Wiley and Sons, New York (1986).

Woodruff, M.L. and M.D. Brownds, *J. Gen. Phsiol.*, **73**, 629 (1979).

Woody, C.D., "Memory, Learning and Higher Functions," Springer-Verlag: Amsterdam (1982).

Yamasaki, Y., M. Fujiwara and N. Toda, *J. Pharmacol. Exp. Ther.*, **190**, 15 (1974).

Yanagihara, K., A. Noma and H. Irisawa, *Jap. J. Physiol.*, **30**, 841 (1980).

Yoshikami, S., J.S. George and W.A. Hagins, *Nature*, **287**, 395 (1980).

AUTHOR INDEX

SUBJECT INDEX

RETURN TO ➡ CHEMISTRY LIBRARY
100 Hildebrand Hall • 642-3753

LOAN PERIOD 1	2	3
4	5	1 MONTH

ALL BOOKS MAY BE RECALLED AFTER 7 DAYS
Renewable by telephone

DUE AS STAMPED BELOW

SEP 24 1999	
APR 18 '01	
DEC 19 2002	
AUG 15 2003	
NOV 04 '04	
DEC 20 2005	